A CONCISE HISTORY OF
MATHEMATICS

BY
DIRK J. STRUIK

Fourth Revised Edition

DOVER PUBLICATIONS
Garden City, New York

For Ruth
and
Rebekka

Bibliographical Note

This Dover edition, first published in 1987, is a further revised and enlarged version of the work originally published by Dover Publications in 1948 and subsequently revised and enlarged in 1967.

Book design by Carol Belanger Grafton.

Library of Congress Cataloging-in-Publication Data

Struik, Dirk Jan, 1894–
    A concise history of mathematics.
    Includes bibliographies and index.
    ISBN-13: 978-0-486-60255-4
    ISBN-10: 0-486-60255-9
    1. Mathematics—History.   I. Title.
QA21.S87   1987   510'.09   86-8855

Manufactured in the United States by LSC Communications Book LLC
60255922   2021
www.doverpublications.com

Preface to the Fourth Revised Edition

This new edition incorporates into the text numerous amendments and corrections, and includes copious additions to the bibliographies. A new chapter has been added on the mathematics of the first half of the present century; in the Preface to the previous edition (dated 1966) I expressed the desire that such a history be written. I now have tried my hand at it myself. The late Professor P. B. Pogrebysskiĭ,* translator of the book into Russian, had treated the subject in an appendix (second edition of translation, Moscow, 1969) which can also be found, translated, in the fifth edition of the German translation (Berlin, 1972). My own approach to this complicated subject has been based in part on Pogrebysskiĭ and on different monographs and chapters in such books as those of C. Boyer, M. Kline, J. M. Dubbey, H. Wussing, and others, but also on my own experience. My version is, again, a "concise" history; there is still room for a full-scale history of mathematics from the beginning of the century to the Second World War.

After the original publication of this book, an edition appeared in Great Britain and translations were published in several languages. In some of these translations, paragraphs or sections were added dealing with the mathematics of the country in which the book appeared. (See "Preface to the Third Revised Edition," below.) A few more translations of the book have appeared since 1966. For the Spanish and Italian translations (by Professors P. Lezama and U. Bottazzini, respectively) I have written prefaces with a sketch of the history of mathematics in Spain and Italy. The Italian translation, moreover, is enriched by a 64-page monograph by Professor Bottazzini on the Italian mathematics of the nineteenth century. The bibliographies in these books, as in some other translations, contain titles of interest to the readers in those particular languages.

The author wishes to express his appreciation to Dr. Neugebauer, whose willingness to read the first chapters of the first edition of this book resulted in several improvements. I express here my appreciation for the generous encouragement I have met, and, in particular, acknowledge my debt to E. J. Dijksterhuis, S. A. Joffe, A. P. Juškevič, R. C. Archibald, K. R. Biermann, and the translators of this book into different languages.

---

*Transliterations of Russian names in the text follow the system of *Mathematical Reviews* and the *Zentralblatt für Mathematik*. As this system is of German origin, "j" should be pronounced as though it were "y" in English; further, the Cyrillic letter elsewhere frequently transliterated as "kh" is given as "h" in this system.

Thanks go to all those colleagues who through reviews or correspondence have made me feel at ease. Thanks also to Dover Publications, Inc., for their willingness to undertake a new edition.

DIRK J. STRUIK

Massachusetts Institute of Technology
1986

Preface to the Third Revised Edition

The first edition of this book appeared in 1948. Since then the reception has been generous, both in this country and abroad, even if occasionally a Russian brow was raised for apparent neglect of Čebyšev, as Scotch or French brows may have been darkened because of apparent lack of respect to the memory of Gregory or Roberval. There have been several translations, and in some of them the translators have added material of special interest to their readers. Thus we find a section on Russian mathematics in the Ukrainian translation (Kiev, 1961) and in the Russian one (Moscow, 1964). I myself, when preparing the Dutch version of the book (Utrecht–Antwerp, 1965), added items of interest to Dutch readers. These editions, as well as the German one (Berlin, 1961; 3rd ed., 1965), also contain bibliographical data of particular interest to the respective countries. When these translations were in progress I used the opportunity to add amendments or corrections to the text, so that the book, through its different editions, has undergone a gradual, although not fundamental, evolution. Much of this has been beneficial in preparing this new English edition.

One day, one of my friends in Peking discovered a Chinese translation (Peking, 1956) which he forwarded to me. The translator of this edition, in his preface, praised the book but objected to its treatment of Chinese mathematics. Since I already had some misgivings, I rewrote the section on this subject. In this edition ancient Chinese mathematics now appears, as it should, as an integral part of medieval and pre-medieval mathematics, and not as a phenomenon outside of the main current of scientific development.

This new edition follows the history of mathematics to the end of the nineteenth century. In 1966 this is quite far away. It is time that the history of mathematics from 1900 to 1950 be written, if only in the form of a "concise history." Nobody seems to have attempted it, although some monographs exist. This is the more remarkable since the market teems with histories of twentieth-century physics. Although this history has the advantage of being more spectacular and (at any rate in some important aspects) more easily understood, the period that began with Poincaré, Hilbert, Lebesgue, Peano, Hardy, and Levi-Civita offers a wealth of material for a fascinating history of mathematics, both in its own right and in relation to logic, physics, and engineering. Who of you, gentle readers, is going to take the initiative?

DIRK J. STRUIK

Massachusetts Institute of Technology
June, 1966

vii

Contents

List of Illustrations

A Concise History Of
Mathematics

Introduction

1

Mathematics is a vast adventure in ideas; its history reflects some of the noblest thoughts of countless generations. It was possible to condense this history into a book of two to three hundred pages only by subjecting ourselves to strict discipline, sketching the unfolding of a few main ideas and minimizing reference to other developments. Bibliographical details had to be restricted to an outline; many relatively important authors—Roberval, Lambert, Schwarz—had to be bypassed. Perhaps the most crippling restriction was the insufficient reference to the general cultural and sociological atmosphere in which the mathematics of a period matured—or was stifled. Mathematics has been influenced by agriculture, commerce, and manufacture, by warfare, engineering, and philosophy, by physics and by astronomy. The influence of hydrodynamics on function theory, of Kantianism and surveying on geometry, of electromagnetism on differential equations, of Cartesianism on mechanics, and of scholasticism on the calculus could only be indicated in a few sentences—or perhaps a few words—yet an understanding of the course and content of mathematics can be reached only if all these determining factors are taken into consideration. Often a reference to the literature has had to replace an historical analysis. Our story ends around 1945, for we feel that the mathematics of the last decades of the twentieth century has so many aspects that it is impossible—to this author at any rate—to do justice even to the main trends.

It is hoped that, despite these restrictions, we have been able to give a fairly honest description of the main trends in the development of mathematics throughout the ages and of the social and cultural setting in which it took place. The selection of the material was, of course, not based exclusively on objective factors, but was influenced by the author's likes and dislikes, his knowledge and his ignorance. As to his ignorance, it was not always possible to consult all sources first-hand; too often, second- or even third-hand sources had to be used. It is therefore good advice, not only with respect to this book, but with respect to all such histories, to check the statements as much as possible with the original sources. This is a good principle for more than one reason. Our knowledge of authors such as Euclid, Diophantus, Descartes, Laplace, Gauss, or Riemann should not be obtained exclusively from quotations or histories describing their works. There is the same invigorating power in the original Euclid or Gauss as there is in the original Shakespeare, and there are places in Archimedes, in Fermat, or in Jacobi which are as beautiful as Horace or Emerson.

Among the principles which have led the author in the presentation of his material are the following:

1. Stress the continuity and affinity of the Oriental civilizations, rather than the mechanical divisions between Egyptian, Babylonian, Chinese, Indian, and Arabian cultures.

2. Distinguish between established fact, hypothesis, and tradition, especially in Greek mathematics.

3. Relate the two trends in Renaissance mathematics, the arithmetical–algebraic and the "fluxional," respectively, to the commercial and engineering interests of the period.

4. Base the exposition of nineteenth-century mathematics on persons and schools rather than on subjects. (Here Felix Klein's history could be used as a primary guide. An exposition by subjects can be found in the books by Cajori and Bell, or with the more technical details, in the *Encyklopædie der mathematischen Wissenschaften* [24 vols., Leipzig, 1898–1935], and in Pascal's *Repertorium der höheren Mathematik* [5 vols., Leipzig, 1910–29].) For the twentieth century a more mixed method has been followed.

2

We list here some of the most important books on the history of mathematics as a whole. Such a list may be taken as a supplement to G. Sarton, *The Study of the History of Mathematics* (Cambridge, 1936), which not only has an interesting introduction to our subject, but also contains full bibliographical information. Consult further K. O. May, *Bibliography and Research Material of the History of Mathematics* (Toronto, 1973; 2nd ed., 1978), 827 pages of bio- and bibliographical information.

English texts to be consulted are:

Archibald, R. C. *Outline of the History of Mathematics*, 6th ed., rev. and enlarged. Math. Assoc. of America, 1949. (An excellent 114-page summary with many bibliographic references.)

Cajori, F. *A History of Mathematics*, 2nd ed. New York, 1938. (A standard text of 514 pp.) First, shorter ed., 1919; Chelsea House reprint, New York, 1980.

Smith, D. E. *History of Mathematics*. 2 vols., Boston, 1923. Dover reprint, 2 vols., 1958. (Restricted mainly to elementary mathematics, but has references concerning all leading mathematicians. Many illustrations.)

Bell, E. T. *Men of Mathematics*. Pelican Books, 1953. Also: E. T. Bell. *The Development of Mathematics*, 2nd ed. New York and London, 1945. (These books contain a wealth of material on mathematicians and their works. The emphasis of the second book is on modern mathematics.)

Scott, J. F. *A History of Mathematics from Antiquity to the Beginning of the Nineteenth Century*. London, 1958.

Eves, H. *An Introduction to the History of Mathematics*. New York, 1953; 4th ed., enlarged, 1976. (Excellent for class use.)

Turnbull, H. W. *The Great Mathematicians*. London, 1929; new ed., New York, 1961. (Also in Newman, J. R. *The World of Mathematics*, Vol. I. New York, 1956.)

Hofmann, J. E. *The History of Mathematics*. New York, 1957. (Trans. from the original German edition, Vol. 1; see reference below. Vols. 2 and 3 of the German edition are translated as *Classical Mathematics*, New York, 1959.)

Boyer, C. B. *A History of Mathematics*. New York, 1968. (With its xv + 717 pages perhaps the best English text for college use. A bibliography of Boyer's work by M. J. Crowe is found in *HM*, Vol. 3 [1976], pp. 397–401.)

Kline, M. *Mathematical Thought from Ancient to Modern Times*. New York, 1972. (This book, with xvii + 1238 pages, has chapters on the history of various subjects, such as ordinary and partial differential equations, abstract algebra, foundations, etc.)

Dealing mainly with elementary mathematics are:

Sanford, V. *A Short History of Mathematics*. London, 1930.

Cajori, F. *A History of Elementary Mathematics*. New York, 1896, 1917.

Ball, W. W. Rouse. *A Short Account of the History of Mathematics*, 6th ed. London, 1915. Dover reprint, 1960. (An older, very readable, but antiquated text; goes no further than the middle of the nineteenth century.)

Bunt, L. N. H.; Jones, P. S.; and Bedient, J. D. *The Historical Roots of Elementary Mathematics*. Englewood Cliffs, N.J., 1976. (Selected topics, discussions, and exercises for school use.)

Historical Topics for the Mathematical Classroom, 31st Yearbook, Nat. Council Teachers of Mathematics. Washington, D.C., 1969. (Monographs on different subjects, each by a special author.)

The standard work on the history of mathematics up to 1800 remains, after all:

Cantor, M. *Vorlesungen über Geschichte der Mathematik*. 4 vols., Leipzig, 1880–1908. (This enormous work, of which the fourth volume was written by a group of specialists under Cantor's direction, covers the history of mathematics until 1799. It is antiquated in many places [especially on Oriental mathematics], but remains a good book for first orientation. Corrections by G. Erneström *et al.* in the volumes of *Bibliotheca mathematica*.)

Other German books are:

Zeuthen, H. G. *Geschichte der Mathematik im Altertum und Mittelalter*. 1st ed., Copenhagen, 1896; French ed., Paris, 1902; 2nd ed. revised by O. Neugebauer, Copenhagen, 1949.

———. *Geschichte der Mathematik im XVI. und XVII. Jahrhundert*. Leipzig, 1903.

Günther, S., and Wieleitner, H. *Geschichte der Mathematik*. 2 vols., Leipzig: Vol. I (by Günther), 1908; Vol. II, 2 parts (by Wieleitner), 1911–21. Ed. by Wieleitner, Berlin, 1939.

Tropfke, J. *Geschichte der Elementar-Mathematik*, 2nd ed. 7 vols., Leipzig, 1921–24. (Vols. I–IV in 3rd ed., 1930–40; Vol. I in 4th ed., revised, Berlin, 1980.)

Die Kultur der Gegenwart. 3 vols., Leipzig and Berlin, 1912. (Contains: Zeuthen, H. G., *Die Mathematik im Altertum und im Mittelalter*; Voss, A., *Die Beziehungen der Mathematik zur allgemeinen Kultur*; Timerding, H. E., *Die Verbreitung mathematischen Wissens und mathematischer Auffassung*.)

Becker, O., and Hofmann, J. E. *Geschichte der Mathematik*. Bonn, 1951.

Hofmann, J. E. *Geschichte der Mathematik*. 3 vols., Berlin, 1953–57. English trans., New York, 1957. (Vol. I, 2nd ed., Berlin, 1963.)

These books contain extensive bio-bibliographies:

Becker, O. *Grundlagen der Mathematik in geschichtlicher Entwicklung*. Freiburg and Munich, 1954; 2nd ed., Freiburg, 1964.

Kowalewski, G. *Grosse Mathematiker*. Munich and Berlin, 1938.

Meschkowski, H. *Problemgeschichte der Mathematik*. 2 vols., Mannheim, 1979, 1980.

———. *Ways of Thought of Great Mathematicians*. San Francisco, 1964.

[Wussing, H.; Arnold, W., ed.] *Biographien bedeutender Mathematiker*. 2nd ed., Berlin, 1978. (Forty-one biographies from Pythagoras to Emmy Noether.)

Wussing, H. *Vorlesungen zur Geschichte der Mathematik*. Berlin, 1979. (The book leads up to recent times, with selected topics such as computers. Strong social background.)

The oldest textbook on the history of mathematics (apart from Proclus) that is more than a catalog is:

Montucla, J.-E. *Histoire des mathématiques*. 4 vols., Paris, 1799–1802. New reprint, 1960. (Deals also with applied mathematics. First published in 1758, 2 vols., but is still good reading.)

Also in French:

d'Ocagne, M. *Histoire abrégée des sciences mathématiques, ouvrage recueilli et achevé par R. Dugas*. Paris, 1952. (Gives short sketches of persons.)

Dedron, J., and Itard, J. *Mathématiques et mathématiciens*. Paris, 1959. (Many illustrations.)

Bourbaki, N. *Eléments d'histoire des mathématiques*. Paris, 1960. (A collection of historical notes from the series *Eléments des mathématiques*, Paris, 1939, till the present.)

[Dieudonné, J., ed.] *Abrégé d'histoire des mathématiques 1700–1900*. 2 vols., Paris, 1970. (A collection of essays by different authors.)

See also M. Daumas and R. Taton, below.

Collette, J. P. *Histoire des mathématiques*. Montreal, 1973. (Leads up to the beginning of the seventeenth century.)

In Italian:

Loria, G. *Storia delle matematiche*. 3 vols., Turin, 1929–33.

Maracchia, S. *La matematica come sistema ipotetico-deduttivo, profile storico*. Florence, 1975.

Frajese, A. *Attraverso la storia della matematica*. Florence, 1973.

In Russian:

Rybnikov, K. A. *Istoriya matematiki*. 2 vols., Moscow, 1960–63.

[Juškevič, A. P., ed.] *Istoriya matematiki*. 3 vols., Moscow, 1970–72. (A collection of essays by different authors.)

There also exist anthologies of mathematical works:

Midonick, H. *A Treasury of Mathematics*. New York, 1965.
Smith, D. E. *A Source Book in Mathematics*. London, 1929.
Wieleitner, H. *Mathematische Quellenbücher*. 4 vols., Berlin, 1927–29.
Speiser, A. *Klassische Stücke der Mathematik*. Zurich and Leipzig, 1925.
Newman, J. R. *The World of Mathematics*. 4 vols., New York, 1956. (An anthology of essays by mathematicians on mathematics.)
Struik, D. J. *A Source Book in Mathematics 1200–1800*. Cambridge, Mass., 1969.

In the same "Source Book" series see also books edited by J. van Heijenoort and G. Birkhoff dealing with the nineteenth and twentieth centuries.

Also useful is:

Callandrier, E. *Célèbres problèmes mathématiques*. Paris, 1949.

There also exist histories of special subjects, of which we must mention the following:

Dickson, L. E. *History of the Theory of Numbers*. 3 vols., Washington, 1919–27.
Muir, T. *The Theory of Determinants in the Historical Order of Development*. 4 vols., London, 1906–23. Supplement, *Contributions to the History of Determinants 1900–20*, London, 1930.
von Braunmühl, A. *Vorlesungen über Geschichte der Trigonometrie*. 2 vols., Leipzig, 1900–03.
Dantzig, T. *Number: The Language of Science*, 3rd ed. New York, 1943; also, London, 1940.
Coolidge, J. L. *A History of Geometrical Methods*. Oxford, 1940.
Loria, G. *Il passato e il presente delle principali teorie geometriche*, 4th ed. Turin, 1931.
——. *Storia della geometria descrittiva dalle origini sino ai giorni nostri*. Milan, 1921.
——. *Curve piani speciali algebriche e trascendenti*. 2 vols., Milan, 1930. German ed., previously published, 2 vols., Leipzig, 1910–11.
Cajori, F. *A History of Mathematical Notations*. 2 vols., Chicago, 1928–29.
Karpinski, L. C. *The History of Arithmetic*. Chicago, 1925. (Bibliography of Karpinski's work by P. S. Jones: *HM*, Vol. 3 [1976], pp. 193–202.)
Walker, H. M. *Studies in the History of Statistical Methods*. Baltimore, 1929.
Reiff, R. *Geschichte der unendlichen Reihen*. Tübingen, 1889.
Todhunter, I. *History of the Progress of the Calculus of Variations During the Nineteenth Century*. Cambridge, 1861.
——. *History of the Mathematical Theory of Probability from the Time of Pascal to That of Laplace*. Cambridge, 1865.
——. *A History of the Mathematical Theories of Attraction and the Figure of the Earth from the Time of Newton to That of Laplace*. 2 vols., London, 1873.
Coolidge, J. L. *The Mathematics of Great Amateurs*. Oxford, 1949.
Archibald, R. C. *Mathematical Table Makers*. New York, 1948.
Dugas, R. *Histoire de la mécanique*. Neuchâtel, 1950. Also see *Mathematical Reviews*, Vol. 14 (1953), pp. 341–43.

6   A CONCISE HISTORY OF MATHEMATICS

Boyer, C. *History of Analytic Geometry*. New York, 1950.
——. *History of the Calculus and Its Conceptual Development*. New York, 1949. Dover reprint, 1959.
Beth, E. W. *Geschiedenis der logica*. The Hague, 1944.
Markuschewitz (Markuševic), A. I. *Skizzen zur Geschichte der analytischen Funktionen*. Berlin, 1955 (trans. from the Russian).
Goldstine, H. H. *A History of the Calculus of Variations from the 17th Through the 19th Century*. New York, etc., 1980.
Dobrovolskiï, V. A. *Essays on the Development of the Analytical Theory of Differential Equations* (Russian). Kiev, before 1976. (See *HM*, Vol. 3 [1976], pp. 221–23.)
Caruccio, E. *Matematica e logica nella storia e nel pensiero contemporaneo*. Turin, 1958. English translation: London, 1964.
Styazhkin, N. I. *History of Mathematical Logic from Leibniz to Peano*. Translated from the Russian: Cambridge, Mass., 1969. (See *HM*, Vol. 2 [1975], pp. 361–65.)
Tietze, H. *Gelöste und ungelöste mathematische Probleme aus alter und neuer Zeit*. Munich, 1949; 2nd ed., Zurich, 1959. English translation: New York, 1965.
Dieudonné, J. *Cours de géométrie analytique*. Paris, 1974. (The first volume is historical.)
Lebesgue, H. *Notices d'histoire des mathématiques*. Geneva, 1959. (Biographical notes of A. T. Vandermonde, C. Jordan, and others.)
Struik, D. J. *The Historiography of Mathematics from Proklos to Cantor*, NTM (Leipzig), Vol. 17 (1980), pp. 1–22.
Maistrov, L. E. *Probability Theory, a Historical Sketch*. New York and London, 1974. (From the Russian, 1967.)
[Grattan-Guinness, 1st ed.] *From the Calculus to Set Theory 1630–1910*. London, 1980. (A collection of essays by different authors.)
Biggs, N. C. *Graph Theory 1736–1936*. Oxford, 1976.
Glaser, A. *History of Binary and Other Non-Decimal Numeration*. Southampton, Pa., 1971.

The history of mathematics is also discussed in the books on the history of science in general. The standard work is:

Sarton, G. *Introduction to the History of Science*. 3 vols., Washington and Baltimore, 1927–48. (This leads up to the fourteenth century and can be supplemented with Sarton's essay, *The Study of the History of Science, with an Introductory Bibliography*, Cambridge, 1936. See also Sarton's book mentioned above and the same author's *History of Science: Ancient Science through the Golden Age of Greece*. Cambridge, Mass., 1952.)

Also:

[Daumas, M., ed.] *Histoire de la science*. Encyclopédie de la Pléiade. Paris, 1957.
[Taton, R., ed.] *Histoire générale des sciences*. Vols. I, II, III$_1$, III$_2$, Paris, 1957–64.

Good texts for school use are:

Sedgwick, W. T., and Tyler, H. W. *A Short History of Science*, 3rd ed. New York, 1948.
Singer, C. *A Short History of Science to the Nineteenth Century*. Oxford, 1941, 1946.

The cultural influence of mathematics and the influence of culture on mathematics are the topics of the following works:

Kline, M. *Mathematics in Western Culture*. New York, 1953.
Bochner, S. *The Role of Mathematics in the Rise of Science*. Princeton, N.J., 1966.
Wilder, R. L. *Evolution of Mathematical Concepts*. New York, 1968. See also *HM*, Vol. 1 (1979), pp. 29–46; Vol. 6 (1979), pp. 57–62.
——. *Mathematics as a Cultural System*. Oxford, etc., 1981.

On the borderline between philosophy and the history of mathematics stands:

[Worrall, J.; Zahar, E., ed.] *Imre Lakatos. Proofs and Refutation. The Logic of Mathematical Discovery*. Cambridge, 1976.

On mathematical discovery, see:

Hadamard, J. *The Psychology of Invention in the Mathematical Field*. Princeton, N.J., 1945, 1949 (from the French, 1939). Dover reprint, 1954.

Also useful are:

Miller, G. A., "A First Lesson in the History of Mathematics," "A Second Lesson," etc. A series of ten articles in *Nat. Math. Mag.*, Vols. 13–19 (1939–45).
[Poggendorff, J. C., ed.] *Biographisch-literarisches Handwörterbuch zur Geschichte der exakten Wissenschaften*. First published in 2 vols., 1863. Continued and extended in the course of the years. By 1974 it comprised 18 volumes; see *Supplement-Band* to *Band VIIa* (Berlin, 1971).
Naas, J.; Schmidt, H. L. *Mathematisches Wörterbuch*. 2 vols., Berlin and Leipzig, 1961.
Meschkowski, H. *Mathematiker Lexikon*, 3e Auflage, Zurich, etc. 1980. (First edition, 1968. Contains short biographies of mathematicians, a bibliography of collected works, and a bibliography of the biographies of mathematicians.)

The reader may be entertained by:

Struik, D. J. "Why Study History of Mathematics?" *UMAP Journal*, Vol. 1 (1980), pp. 3–28.

The following periodicals deal with the history of mathematics (or of science in general):

Archiv für die Geschichte der Mathematik, 1909–31.
Bibliotheca mathematica, Ser. 1–3, 1884–1914.
Quellen und Studien zur Geschichte der Mathematik, 1931–38.
Scripta mathematica, 1932–present.
Isis, 1913–present.
Revue d'histoire des sciences, 1947–present.
Archives internationales d'histoire des sciences, 1947–present (former *Archeion*).
Annals of Science, 1936–present.
Scientia, 1907–present.
Centaurus, 1950–present.

Istoriko-matematičeskie Issledovaniya, 1948–present.
Boethius, 1962–present.
Physis, 1959–present.
NTM (*Zeitschrift für Geschichte der Naturwissenschaften, Technik und Medizin*), 1960–present.
AHES (*Archive for History of Exact Sciences*), 1960–present.
Biometrika, 1901–present.
HM (*Historia Mathematica*), 1974–present.
Indian Journal of History of Science.
Bollettino di Storia delle Scienze Matematiche, 1981–present.
Journal of the History of Arabic Science.
Annals of the History of Computers, 1979–present.

AHES and *HM* are exclusively devoted to mathematics. *HM* keeps the bibliography up to date, as does *Isis* with the whole of the history of science. See also *Mathematical Reviews*, 1940–present, with a section on history, and its German and Russian counterparts.

Biographies of outstanding mathematicians (deceased) are found in the fifteen volumes of the *Dictionary of Scientific Biography* (*DSB*), New York, 1970–80. (Vol. 15 is the index.)

A World Directory of Historians of Mathematics, 2nd ed., 1981, has been prepared by the Institute for the History and Philosophy of Science and Technology at the University of Toronto, Canada. Later edition prepared by Prof. L. Nový, at the Czechoslovakian Academy of Sciences in Prague.

CHAPTER I

The Beginnings

1

Our first conceptions of number and form date back to times as far removed as the Old Stone Age, the Paleolithic. Throughout the hundreds or more millennia of this period men lived in small groups, under conditions differing little from those of animals, and their main energies were directed toward the elementary process of collecting food wherever they could get it. They made weapons for hunting and fishing, developed a language to communicate with each other, and in later paleolithic times enriched their lives with creative art forms, including statuettes and paintings. The paintings in caves of France and Spain (over 15,000 years old) may have had some ritual significance; certainly they reveal a remarkable understanding of form; mathematically speaking, they reveal understanding of two-dimensional mapping of objects in space.

Little progress was made in understanding numerical values and space relations until the transition occurred from the mere *gathering* of food to its actual *production*, from hunting and fishing to agriculture. With this fundamental change, a revolution in which the passive attitude of man toward nature turned into an active one, we enter the New Stone Age, the Neolithic.

This great event in the history of mankind occurred perhaps ten thousand years ago, after the ice sheet that covered Europe and Asia had melted and made room for forests and deserts. Here nomadic wandering in search of food came slowly to an end. Fishermen and hunters were in large part replaced by simple farmers. Such farmers, remaining in one place as long as the soil stayed fertile, began to build more permanent dwellings; villages emerged as protection against the climate and against predatory enemies. Many such neolithic settlements have been excavated. The remains show how gradually elementary crafts such as pottery, carpentry, and weaving developed. There were granaries, so that the inhabitants were able to provide against winter and hard times by establishing a surplus. Bread was baked, beer was brewed, and in late neolithic times copper and bronze were smelted and prepared. Inventions appeared, notably the potter's wheel and the wagon wheel; boats and shelters were improved. All these remarkable innovations occurred only within limited areas and did not always spread to other localities. The American Indian, for example, did not learn much about the technical use of the wagon wheel until the coming of the European. Nevertheless, as compared with paleolithic times, the tempo of technical improvement was enormously accelerated.

Between the villages a considerable trade existed, which so expanded that connections can be traced between places hundreds of miles apart. The discovery of the arts of smelting and manufacturing, first copper, then bronze tools and weapons, strongly stimulated this commercial activity. This again promoted the further formation of languages. The words of these languages expressed very concrete things and very few abstractions, but already there was room for some simple numerical terms and for some form relations. Many Australian, American, and African tribes were in this stage at the period of their first contact with Europeans: some tribes are still living in these conditions, so that it is possible to study their habits and forms of expression, and to some extent to understand them if we can strip ourselves of preconceived notions.

2

Numerical terms—expressing some of "the most abstract ideas which the human mind is capable of forming," as Adam Smith has said—came only slowly into use. Their first occurrence was qualitative rather than quantitative, making a distinction only between one (or better "a"—"a man," rather than "one man") and two and many. In the old Fiji Island language ten boats are called *bola*, ten coconuts *koro*, and a thousand coconuts *saloro*. The ancient qualitative origin of numerical conceptions can still be detected in the special dual terms existing in certain languages such as Greek or Celtic. When the number concept was extended, higher numbers were first formed by addition: 3 by adding 2 and 1, 4 by adding 2 and 2, 5 by adding 2 and 3.

Here is an example from some Australian peoples:

Murray River: 1 = enea, 2 = petcheval, 3 = petcheval-enea, 4 = petcheval petcheval

Kamilaroi: 1 = mal, 2 = bulan, 3 = guliba, 4 = bulan bulan, 5 = bulan guliba, 6 = guliba guliba.[1]

The development of the crafts of commerce stimulated this crystallization of the number concept. Numbers were arranged and bundled into larger units, usually by the use of the fingers of the hand or of both hands, a natural procedure in trading. This led to numeration first with five, later with ten as a base, completed by addition and sometimes by subtraction, so that 12 was conceived as 10 + 2, or 9 as 10 − 1. Sometimes 20, the number of fingers and toes, was selected as a base. Of 307 number systems of primitive American peoples investigated by W. C. Eels, 146 were decimal, 106 quinary and quinary decimal, vigesimal and quinary vigesimal.[2] The vigesimal system in its most characteristic form occurred among the Mayas of Mexico and the Celts in Europe.

---

[1] L. Conant, *The Number Concept* (New York, 1896), pp. 106–7, with many similar examples.

[2] W. C. Eels, "Number Systems of North American Indians," *Amer. Math. Monthly*, Vol. 20 (1913), pp. 263–72, 293–99; see esp. p. 293.

Numerical records were kept by means of bundling: strokes on a stick, knots on a string, pebbles or shells arranged in heaps of fives—devices very much like those of the old-time innkeeper with his tally stick. From this method to the introduction of special symbols for 5, 10, 20, etc., was only a step, and we find exactly such symbols in use at the beginning of written history, at the so-called dawn of civilization.

One of the oldest examples of the use of a tally stick dates back to paleolithic times and was found in 1937 in Věstonice (Moravia). It is the bone of a young wolf, 7 inches long, engraved with 55 deeply incised notches, of which the first 25 are arranged in groups of 5. They are followed by a simple notch twice as long which terminates the series; then, starting from the next notch, also twice as long, a new series runs up to 30.[3] Other such marked sticks have been found.

It is therefore clear that the old saying found in Jakob Grimm and often repeated, that "counting started as finger counting," is incorrect. Counting by fingers, that is, counting by fives and tens, came only at a certain stage of social development. Once it was reached, numbers could be expressed with reference to a base, with the aid of which large numbers could be formed; thus a primitive type of arithmetic originated. Fourteen was expressed as $10 + 4$, sometimes as $15 - 1$. Multiplication began where 20 was expressed not as $10 + 10$, but as 2×10. Such dyadic operations were used for millennia as a kind of middle road between addition and multiplication, notably in Egypt and in the pre-Aryan civilization of Mohenjo-Daro on the Indus. Division began where 10 was expressed as "half of a body," although conscious formation of fractions remained extremely rare. Among North American tribes, for instance, only a few instances of such formations are known, and this is in almost all cases only of 1/2, although sometimes also of 1/3 or 1/4.[4] A curious phenomenon was the love of very large numbers, a love perhaps stimulated by the all-too-human desire to exaggerate the extent of herds of enemies slain; remnants of this tendency appear in the Bible and in other sacred and not-so-sacred writings.

3

It also became necessary to measure the length and contents of objects. The standards were rough and often taken from parts of the human body, and in this way units such as fingers, feet, or hands originated. The names "ell," "fathom," and "cubit" remind us also of this custom. When houses were built, as among the agricultural Indians or the pole-house dwellers of Central Europe, rules were laid down for building along straight lines and at right angles. The word "straight" is related to "stretch," indicating operations with a rope;[5] the word "line" to "linen,"

---

[3]*Isis*, Vol. 28 (1938), pp. 462–63 (from *Illustrated London News*, Oct. 2, 1937).

[4]G. A. Miller has remarked that the words *one-half*, *semis*, *moitié* have no direct connection with the words *two*, *duo*, *deux* (contrary to *one-third*, *one-fourth*, etc.), which seems to show that the conception of 1/2 originated independent of that of integer. *Nat. Math. Mag.*, Vol. 13 (1939), p. 272.

[5]The name "rope-stretchers" (Greek: *harpedonaptai*; Arabic: *massah*; Assyrian: *masihānu*) was attached in many countries to men engaged in surveying—see S. Gandz, in *Quellen und Studien zur Geschichte der Mathematik*, Vol. I (1931), pp. 255–77.

GEOMETRICAL PATTERNS DEVELOPED BY AMERICAN INDIANS.
(From Spier, see "Literature," below.)

showing the connection between the craft of weaving and the beginnings of geometry.[6] This was one way in which interest in mensuration evolved.

Neolithic man also developed a keen feeling for geometrical patterns. The baking and coloring of pottery, the plaiting of rushes, the weaving of baskets and textiles, and later the working of metals led to the cultivation of plane and spatial relationships. Dance patterns must also have played a role. Neolithic ornamentation rejoiced in the revelation of congruence, symmetry, and similarity. Numerical relationships might enter into these figures, as in certain prehistoric patterns which represent triangular numbers; others display "sacred" numbers.

Figures 1–4 below give examples of some interesting geometrical patterns occurring in pottery, weaving, and basketry. The design in Fig. 1 can be found on

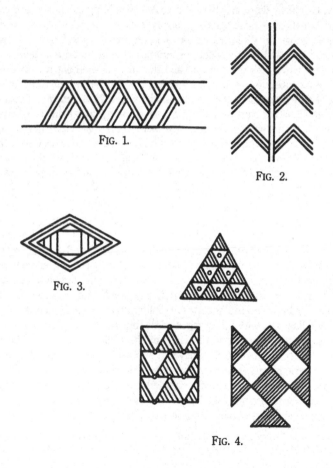

FIG. 1.

FIG. 2.

FIG. 3.

FIG. 4.

---

[6]For the derivation of the names for numbers, see the book by K. Menninger (see "Literature," below).

neolithic pottery in Bosnia and on objects of art in the Mesopotamian Ur period.[7] The motif in Fig. 2 exists on Egyptian pottery of the Predynastic period (c. 4000–3500 B.C.).[8] Fig. 3 shows patterns which were used by pole-house dwellers near Ljubljana (Yugoslavia) in the Hallstatt period (Central Europe, c. 1000–500 B.C.).[9] The designs in Fig. 4, rectangles filled with triangles, triangles filled with circles, are from urns in graves near Sopron in Hungary. They show attempts at the formation of triangular numbers, which played an important role in Pythagorean mathematics of a later period.[10]

Patterns of this kind have remained popular throughout historical times. Beautiful examples can be found on dipylon vases of the Minoan and early Greek periods, in the later Byzantine and Arabian mosaics, and on Persian and Chinese tapestry. Originally there may have been a religious or magic meaning to the early patterns, but their esthetic appeal gradually became dominant. In the religions of the Stone Age we can discern attempts at conforming to the forces of nature, the social structure, and the individual experience. Religious ceremonies were permeated with what we see as magic, and this magical element was incorporated into existing conceptions of number and form as well as in sculpture, music, and drawing. There were magical numbers (such as 3, 4, 7) and magical figures (such as the Pentalpha and the Swastika). Some authors have even considered this aspect of mathematics the determining factor in its growth,[11] but though the social roots of mathematics may have become obscured in modern times, they are fairly obvious during this early period of man's history. "Modern" numerology is a leftover from magical rites dating back to neolithic, and perhaps even to paleolithic, times.

4

Even among peoples with a social structure far removed from our technical civilization we find some reckoning of time, and, closely related, some knowledge of the motion of sun, moon, and stars. This knowledge first attained its more scientific character when farming and trade expanded. The use of a lunar calendar dates very far back into the history of mankind, the changing aspects of vegetation being connected with the changes of the moon. People at an early date also paid attention to the solstices or rising of the Pleiades at dawn. The earliest people with records attributed a knowledge of astronomy to their most remote, prehistoric periods. Other peoples used the constellations as guides in navigation. From this astronomy resulted some knowledge of the properties of the sphere, of angular directions, of circles, and of even more intricate figures.

In recent years, considerable attention has been paid to the possible astronomi-

[7]W. Lietzmann, "Geometrie und Praehistorie," *Isis*, Vol. 20 (1933), pp. 436–39.

[8]D. E. Smith, *History of Mathematics* (Boston, 1923), Vol. I, p. 15. (Dover reprint, 2 vols., 1958.)

[9]M. Hoernes, *Urgeschichte der bildenden Kunst in Europa* (Vienna, 1915).

[10]See also F. Boas, *General Anthropology* (New York, 1938), p. 273.

[11]W. J. McGee, *Primitive Numbers*, Nineteenth Annual Report, Bureau Amer. Ethnology 1897–98 (1900), pp. 825–51. See esp. the papers by A. Seidenberg, cited below.

cal and calendrical significance of prehistoric stone monuments, such as Stone-henge in England, dating back to c. 2000 B.C.[12] If they had such significance, then the question arises whether this astronomical, and thus also mathematical, knowl-edge was transmitted from some center, perhaps Mesopotamia (diffusionism), or had a native, autochthonous origin (an opinion that has gained much support lately). The civilizations in America seem to have developed either independently from those of Eurasia and Africa, or at any rate with little interference from them.

5

These few illustrations of the beginnings of mathematics show that the historical growth of a science does not necessarily pass through the stages in which we now develop it in our instruction. Some of the oldest geometrical forms known to mankind, such as knots and patterns, only received full scientific attention in recent years. On the other hand, some of our more elementary branches of mathematics, such as the graphical representation or elementary statistics, date back to comparatively modern times. As A. Speiser has remarked with some asperity (and some exaggeration):

> Already the pronounced tendency toward tediousness, which seems to be inherent in elementary mathematics, might plead for its late origin, since the creative mathematician would prefer to pay his attention to the interesting and beautiful problems.[13]

6

This may be a good place to mention the mathematics of three ancient civiliza-tions, interesting in themselves, but of little or no influence on the further course of mathematics: the Minoans-Mycenaeans, the Mayas, and the Incas. Their science is not that of "the beginnings," but belongs rather to the category of the next chapter, on the Ancient Orient.

Mathematical symbols used in administration have been found in the ruins of the Minoan-Mycenaean civilization of Crete and the Greek mainland. They belong to the scripts called Linear A and B and belong to the period of c. 1800–1200 B.C. Numbers are represented, as in Egypt (but with different symbols), by special symbols for 1, 10, 100, 1000 in an additive way. There are also symbols for simple fractions, not all unit fractions. Since the scribes did not bake the clay tablets on which they wrote, only those that were baked in the final conflagration of their cities have been preserved, so that we have inadequate knowledge of the extent of the mathematical knowledge of this civilization; it may have been comparable to that of Egypt. At any rate we know that Homer's heroes had scribes who could do some arithmetic on tablets.

The Mayas of Central America, mainly in what is now Yucatán and Guatemala,

---

[12]For a general discussion of this subject see the book by D. C. Haggie, cited below.

[13]This is a witty remark, but "elementary mathematics," taught by a good instructor, need not be tedious at all. And do not the regular polyhedra and the golden section, which have excited people from Plato to the present, belong to "elementary mathematics"?

established a civilization that lasted for a millennium and a half, but reached its height in the so-called classical period, about 200–900 of our era. The arithmetic of the Mayas, mainly deciphered from inscribed stone monuments, some codices, and Spanish chronicles, and closely related to their astronomy, notably their calendric system, was vigesimal (it still is), represented by dots for the units up to 4, and horizontal bars for the fives up to 15. For larger numbers they used a position system with base 20, powers of 20 being represented by the same symbol as 20, the unit symbol. There were some modifications for calendric purposes. This position system required a symbol for zero, often a kind of shell or half-open-eye sign. This system, with its calendric connections, spread to other peoples of Central America. We think of the famous calendar stone found in Mexico City, dating from the time of the Aztecs, who came to this location at the end of the eleventh century of our era.

The Incas built a large empire in and west of the Andes of South America from the middle of the thirteenth century of our era on, their capital being Cuzco. Its vast bureaucracy, strong in administration, crafts, and engineering, used, for communication and information, no writing, but so-called *quipos*. The simplest quipo has a main cord of colored cotton or sometimes wool, from which knotted cords are suspended with the knots formed into clusters at some distance from each other. Each cluster has a number of knots from 1 to 9, and a cluster of, say, 4 followed by one of 2 and one of 8 knots represents 428. This is therefore a position system, in which our zero is indicated by a greater distance between the knots. The colors of the cords represent things: sheep, soldiers, etc.; and the position of the cords, as well as additional cords suspended from the cords, could tell a very complicated statistical story to the scribes who could "read" the quipos.

Quipos may have hundreds of knotted cords; the largest so far found has 1,800 pendent cords; it may have indicated the composition of an army or work force. Only some 400 quipos have been discovered, all in graves, since the Spanish destroyed quipos as ungodly.

These quipos teach us that we can have a sophisticated bureaucratic civilization without the art of writing. Is it possible that cultures like that of Stonehenge had similar means of communication and storage of information that are now gone forever? Those quipos that survived were buried in the desert region along the Pacific; those buried in less arid regions have all been lost.

Literature

Apart from the texts by Conant, Eels, Smith, Lietzmann, McGee, and Speiser already quoted, see:

Menninger, K. *Zahlwort und Ziffer, Eine Kulturgeschichte der Zahlen*, 2nd ed. 2 vols., Göttingen, 1957–58.
Struik, D. J. "Stone Age Mathematics." *Scientific American*, Vol. 179 (December, 1948), pp. 44–49.

PREDYNASTIC EGYPTIAN VASE.
*Courtesy of the Metropolitan Museum of Art
(Gift of Mrs. Helen Miller Gould, 1910).*

mith, D. E., and Ginsburg, J. *Numbers and Numerals*. New York Teachers' College, 1937.
Childe, Gordon. *What Happened in History*. Pelican Books, 1942.

Interesting patterns are described in:

Spier, L. "Plains Indian Parfleche Designs." *Univ. of Washington Publ. in Anthropology*, Vol. 4 (1931), pp. 293–322.
Deacon, A. B. "Geometrical Drawings from Malekula and Other Islands of the New Hebrides," *J. Roy. Anthrop. Institute*, Vol. 64 (1934), pp. 129–75.
Popova, M. "La géométrie dans la broderie bulgare." *Comptes Rendus, Premier Congrès des Mathématiciens des pays slaves* (Warsaw, 1929), pp. 367–69.

On the mathematics of the American Indians, see also:

Thompson, J. E. S. "Maya Arithmetic." *Contributions to Amer. Anthrop. and History*, Vol. 36, Carnegie Inst. of Washington Publ. No. 528 (1941), pp. 37–62.
Smith, D. E. *History of Mathematics*. Boston, 1923. Dover reprint, 2 vols. in 1, 1958. (See the extensive bibliography on p. 14.)
Lounsbury, F. C. *Maya Numeration, Computation, and Calendrical Astronomy, DSB*, Vol. 15 (1978), pp. 759–818. (Detailed study with extensive bibliography.)
Ascher, M. and D. *Code of the Quipos: A Study in Media, Mathematics and Culture*. Ann Arbor, Mich., 1981. See also *AHES*, Vol. 8 (1972), pp. 288–320, and *Visible Language* (Cleveland, Ohio, 1975), pp. 329–56.

On African mathematics:

Zaslavsky, C. *Africa Counts*. Boston, 1973.
Crowe, D. W. "The Geometry of African Art." I, *Journal of Geometry*, Vol. 1 (1971), pp. 169–82; II, *HM*, Vol. 2 (1975), pp. 253–71.

On "megalithic" astronomy and mathematics, e.g. Stonehenge:

Hawkins, G. S. *Beyond Stonehenge*. London, 1973.
Haggie, D. C. *Megalithic Science*. London, 1981.

On the relation of ritual to mathematics:

Seidenberg, A. "The Ritual Origin of Geometry," *AHES*, Vol. 1 (1960–61), 480–527.
———. "The Ritual Origin of Counting," *AHES*, Vol. 2 (1962), pp. 1–40.
———. "The Origin of Mathematics," *AHES*, Vol. 18 (1970), pp. 301–42.

For the development of mathematical concepts in children, see:

Riess, A. *Number Readiness in Research*. Chicago, 1947.
Piaget, J. *La genèse du nombre chez l'enfant*. Neuchâtel, 1941.
———. *Le développement des quantités chez l'enfant*. Neuchâtel, 1941.
Bunt, L. N. H. *The Development of the Ideas of Numbers and Quantity According to Piaget*. Groningen, 1951.
Freudenthal, H. *Mathematics as an Educational Task*. Dordrecht, 1973.

CHAPTER II

The Ancient Orient

1

During the fifth, fourth, and third millennia B.C. newer and more technically advanced forms of society evolved from well-established neolithic communities along the banks of great rivers in Africa and Asia, in subtropic or nearly subtropic regions. These rivers were the Nile, the Tigris and Euphrates, the Indus and later the Ganges, the Huang He and later the Yangzi. Some societies in America may also reach back to these times.

The lands along these rivers could be made to yield abundant crops once the flood waters were brought under control and the swamps drained. In contrast with the arid desert and mountain regions and plains surrounding these countries, the river valleys could be made into lands of great fertility. Within the course of centuries, these problems were solved by the building of levees and dams, the digging of canals, and the construction of reservoirs. Regulation of the water supply required coordination of activities between widely separated localities on a scale greatly surpassing all previous efforts. This led to the establishment of central organs of administration, located in urban centers rather than in the barbarian villages of former periods. The relatively large surplus yielded by the vastly improved and intensive agriculture raised the standard of living for the population as a whole, but it also created an urban aristocracy headed by powerful chieftains. There were many specialized crafts carried on by artisans, soldiers, clerks, and priests. Administration of the public works was placed in the hands of a permanent officialdom, a group wise in the behavior of the seasons, the motions of the heavenly bodies, the art of land division, the storage of food, and the raising of taxes. A script was used to codify the requirements of the administration and the actions of the chieftains. Bureaucrats as well as artisans acquired a considerable amount of technical knowledge, including metallurgy and medicine. To this knowledge belonged also the arts of computation and mensuration.

By this time social classes were firmly established. There were chieftains, free and tenant farmers, craftsmen, scribes, officials, serfs, and slaves. Local chiefs increased in wealth and power to such an extent that they rose from feudal lords of limited authority to become local kings of absolute sovereignty. Quarrels and wars among the various despots led to larger domains, united under a single monarch. These forms of society based on irrigation and intensive agriculture led in this way to an "Oriental" form of despotism. Such despotism could be maintained for centuries and then collapse, sometimes under the impact of mountain and desert tribes attracted by the wealth of the valleys, or through neglect of the vast, complicated, and vital irrigation system. Under such circumstances, power might shift from one tribal king to another, or society might break up into smaller feudal units and the process of unification would begin all over again. However, under all these dynastic revolutions and recurrent transitions from feudalism to absolutism, the villages, which were the basis of this society, remained essentially unchanged, and with them the fundamental economic and social structure. Oriental society has moved in cycles; even at present there exist many communities in Asia and Africa which have persisted for several millennia in the same pattern of life. Progress under such conditions was slow and erratic, and periods of cultural growth might be separated by many centuries of stagnation and decay.

We must be careful not to ascribe the character of Oriental society too exclusively to its basis in irrigation and intensive agriculture. Questions of land ownership, feudal relations, age-long family traditions played their parts in different ways in different countries of Asia, Africa, and America. But nevertheless underneath there remained a kinship among these various societies, and this is a key to the understanding of the type of mathematics one sees to have been established in them all: an arithmetical-algebraic one.

The static character of the "Orient" (including early American societies) imparted a fundamental sanctity to its institutions which facilitated the identification of religion with the state apparatus. The bureaucracy often shared this religious character of the state; in many Oriental countries priests were the administrators of the domain. Since the cultivation of science was the task of the bureaucracy, we find in many—but not all—Oriental countries that the priests were the outstanding carriers of scientific knowledge.

2

"Oriental" mathematics originated as a practical science in order to facilitate computation of the calendar, administration of the harvest, organization of public works, and collection of taxes. The initial emphasis naturally was on practical arithmetic and mensuration. However, a science cultivated for centuries by a special craft, whose task it is not only to apply it but also to guard its secrets, develops tendencies toward abstraction. Gradually it will come to be studied for its own sake. Arithmetic evolved into algebra not only because it allowed better practical computations, but also because it was the natural outgrowth of a science cultivated and developed in schools of scribes. For these same reasons, mensuration developed into the beginnings—but no more—of a theoretical geometry.

Despite all the trade and commerce in which these ancient societies indulged, their economic core was agricultural, centered in the villages, characterized by isolation and traditionalism. The result was that, despite similarity in economic structure and in the essentials of scientific lore, there always remained striking differences between the different cultures. The seclusion of the Chinese in certain periods and of the Egyptians was proverbial. It always has been easy to differentiate between the arts and script of the Egyptians, the Mesopotamians, the Chinese, and the Indians. In the same way we can speak of Egyptian, Mesopotamian, Chinese, and Indian mathematics, although their general arithmetical-algebraic nature was very much alike. Even if the science of one country progressed beyond that of another during some period, it preserved its characteristic approach and symbolism.

It is difficult to date new discoveries in the East. The static character of its social structure has tended to preserve scientific lore throughout centuries or even millennia. Discoveries made within the seclusion of a township may never have spread to other localities. Storages of scientific and technical knowledge have been destroyed by dynastic changes, wars, or floods. There is a story that in 221 B.C., when China was united under one absolute despot, Shi Huangdi (Shih Huang-ti), he ordered all books of learning to be destroyed. Later many of them were rewritten partly from memory, but such events make the dating of discoveries very difficult.

Another difficulty in dating Oriental science is due to the material used for its preservation. The Mesopotamian people baked clay tablets which are virtually indestructible.[1] The Egyptians used papyrus, and a sizable body of their writing has been preserved in the dry climate. The Chinese and Indians used far more perishable material, such as bark or bamboo. The Chinese, about the first century B.C., began to use paper, but little has been preserved which dates back to the millennia before A.D. 700. Our knowledge of Oriental mathematics is therefore very sketchy. For the pre-Hellenistic centuries we are almost exclusively confined to Mesopotamian and Egyptian material. However, it is entirely possible that new discoveries will lead to a complete reevaluation of the relative merits of the different Oriental forms of mathematics. For a long time our richest historical field lay in Egypt because of the discovery in 1858 of the so-called *Papyrus Rhind*, written about 1650 B.C., but which contained much older material.[2] In the last few decades our knowledge of Babylonian mathematics has been vastly augmented by the remarkable discoveries of O. Neugebauer and F. Thureau-Dangin, who deciphered a large number of clay tablets. It is now apparent that Babylonian mathematics was far more developed than its Oriental counterparts. This judgment may be final, since there exists a certain consistency in the factual character of the Babylonian and Egyptian texts throughout the centuries. Moreover, the economic development of Mesopotamia was more advanced than that of other countries in the so-called Fertile Crescent of the Near East, which stretched from Mesopotamia to Egypt. Mesopotamia was the crossroads for a large number of caravan routes, while Egypt remained in comparative isolation. In addition, the harnessing of the erratic Tigris and Euphrates required more engineering skill

---

[1]Except when they are not carefully preserved after excavation. The loss of tablets by careless handling has been considerable.

[2]Named after the Scottish banker and antiquary A. Henry Rhind (1833–1863), who bought it at Luxor on the Nile. It is in the British Museum. It is also called the Ahmes Papyrus, after the scribe who copied it—the earliest personal name known to us in the history of mathematics.

and administration than that of the Nile, that "most gentlemanly of all rivers" (to quote Sir William Willcocks).[3] Further study of ancient Hindu mathematics may still reveal unexpected excellence, although so far claims for it have not been very convincing.

3

Most of our knowledge of Egyptian mathematics is derived from two mathematical papyri: the *Papyrus Rhind* (already mentioned), containing 85 problems; and the so-called *Moscow Papyrus*, perhaps two centuries older, containing 25 problems. These problems were already ancient lore when the manuscripts were compiled, but there are minor papyri of much more recent date—even from Roman times—which show no difference in approach. The mathematics they profess is based on a decimal system of numeration with special signs for each higher decimal unit—a system with which we are familiar through the Roman system which follows the same principle: MDCCCLXXVIII = 1878. On the basis of this system the Egyptians developed an arithmetic of a predominantly additive character, which means that its main tendency was to reduce all multiplication to repeated additions. For instance, multiplication of 13 by a number was obtained by multiplying the number first by 1, then, using successive doubling, by 2, then by 4, then by 8, and finally adding the results of the multiplication by 1, 4, and 8 (the components of 13).

For example, for the computation of 13 × 11:

$$\begin{array}{ll} *1 & 11 \\ 2 & 22 \\ *4 & 44 \\ *8 & 88 \end{array}$$

Add the numbers indicated by *, which gives 143.

The most remarkable aspect of Egyptian arithmetic was its calculus of fractions. All fractions were reduced to sums of so-called unit fractions, meaning fractions with 1 as numerator. They were indicated by the number of the denominator with a symbol above, which we shall indicate by a bar, so that we shall indicate 1/10 by $\overline{10}$. The only exceptions were 1/2 and 2/3, for which there were special symbols. The reduction to sums of unit fractions was made possible by tables, which gave the decomposition for fractions of the form $2/n$—the only decomposition necessary because of the dyadic multiplication. The *Papyrus Rhind* has a table giving the equivalents in unit fractions for all odd n from 5 to 101, e.g.:

$$\begin{array}{r|lll} n = 5 & \overline{3} & \overline{15} & (2/5 = 1/3 + 1/15) \\ 7 & \overline{4} & \overline{28} \\ 9 & \overline{6} & \overline{18} \\ 59 & \overline{36} & \overline{236} & \overline{531} \\ 97 & \overline{56} & \overline{679} & \overline{776}. \end{array}$$

The principle underlying the special reduction to unit fractions is not clear

---

[3]W. Willcocks, *Irrigation of Mesopotamia*, 2d. ed. (London, 1917), p. xi.

(e.g., why for $n = 19$ the reduction $\overline{12}\ \overline{76}\ \overline{114}$ and not $\overline{12}\ \overline{57}\ \overline{228}$?). This calculus with fractions gave to Egyptian mathematics an elaborate and ponderous character, but despite the handicaps of this work with unit fractions it was practiced for thousands of years, not only during the Greek period, but even during the Middle Ages. The decomposition presupposed some mathematical skill, and there exist interesting theories to explain the way in which the Egyptian specialists might have obtained their results.[4]

Many problems were very simple and did not go beyond a linear equation with one unknown:

A quantity, its 2/3, its 1/2, and its 1/7, added together, gives 33. What is this quantity?

The answer, 14 28/97, is written in unit fractions:

$$14\ \overline{4}\ \overline{97}\ \overline{56}\ \overline{679}\ \overline{776}\ \overline{194}\ \overline{388}.$$

For the unknown in an equation there existed a hieroglyph meaning "heap"; e.g., *hau* or *aha*. Egyptian algebra therefore is sometimes called "*aha*-calculus."

The problems deal with the strength of bread and of different kinds of beer, with the feeding of animals and the storage of grain, showing the practical origin of this cumbersome arithmetic and primitive algebra. Some problems show a theoretical interest, as in the problem of dividing 100 loaves among 5 men in such a way that the share received shall be in an arithmetical progression, and that 1/7 of the sum of the largest 3 shares shall be equal to the sum of the smallest 2. We even find a geometrical progression dealing with 7 houses in each of which there are 7 cats, each cat watching 7 mice, etc., which reveals a knowledge of the formula of the sum of a geometrical progression.[5]

Some problems were of a geometrical nature, dealing mostly with mensuration. The area of the triangle was found to be half the product of base and altitude; the area of a circle with diameter d was given as $\left(d - \dfrac{d}{9}\right)^2$, which led to a value of π of $256/81 = 3.1605$. In addition there are formulas for solid volumes, such as the cube, the parallelepiped, and the circular cylinder, all conceived concretely as containers, mainly of grain. The most remarkable result of Egyptian mensuration was the formula for the value of the frustum of a square pyramid $V = (h/3)(a^2 + ab + b^2)$, where a and b are the lengths of the sides of the squares and h is the height. This result, the counterpart of which has not yet been found in other

---

[4]O. Neugebauer, "Arithmetik und Rechentechnik der Ägypter," *Quellen und Studien zur Geschichte der Mathematik*, Vol. B 1 (1931), pp. 301–80; B. L. van der Waerden, *Die Entstehungsgeschichte der ägyptischen Bruchrechnung*, Vol. 4 (1938), pp. 359–82; E. M. Bruins, "Ancient Egyptian Arithmetic: 2/N," *Proc. Nederl. Akad. Wet.*, Vol. A 55 (1952), pp. 81–91.

[5]We think of the nursery rhyme:

"As I was going to St. Ives
I met a man with seven wives.
Every wife had seven sacks,
Every sack had seven cats,
Every cat had seven kits.
Kits, cats, sacks, and wives,
How many were going to St. Ives?"

We see how the same kind of problem can maintain itself throughout the ages.

A SINGLE PAGE OF THE GREAT "PAPYRUS RHIND."
(From Chace, Vol. II, p. 56).

1
2
3
4
5
6
7
8
9
10
11

oi　m　f·nptb　f·nh f·4 f·ż　　ḥʿc

ż　8ż　4　　　4　żi 1　1

1　　4i　ż　　　ż　3　2

　　　　　　　7　　4

4i　ɣ　ż　5　ḥʿc　3p　dmd　4　ɣ

forms of ancient mathematics, is even more remarkable since there is little indication that the Egyptians had any notion even of the Pythagorean theorem, despite some unfounded stories about *harpedonaptai*, who supposedly constructed right triangles with the aid of a string with $3 + 4 + 5 = 12$ knots.[6]

Here we must warn against exaggerations concerning the antiquity of Egyptian mathematical knowledge. All kinds of advanced science have been credited to the pyramid builders of 3000 B.C. and earlier years; there is even a widely accepted story that the Egyptians of 4212 B.C. adopted the so-called Sothic cycle for the measurement of the calendar. Such precise mathematical and astronomical work cannot be seriously ascribed to a people slowly emerging from neolithic conditions, and the source of these tales can usually be traced to a late Egyptian tradition transmitted to us by the Greeks. It is a common characteristic of ancient civilizations to date fundamental knowledge back to very early times. All available texts point to an Egyptian mathematics of rather limited scope, though quite sophisticated within this scope. Their astronomy was on the same general level. However, with our increasing respect for the astronomical knowledge of people living under Stone Age conditions or just emerging from them, our opinions on these matters may change.

4

Mesopotamian mathematics reached a far higher level than Egyptian mathematics ever obtained. Here we can even detect progress in the course of the centuries. Already the oldest texts, dating from the third millennium to the latest Sumerian period (the third dynasty of Ur, *c.* 2100 B.C.), show keen computational ability. These texts contain multiplication tables in which a well-developed sexagesimal system of numeration was superimposed on an original decimal system; there are cuneiform symbols indicating 1, 60, 3600, and also 60^{-1}, 60^{-2}. However, this was not their most characteristic feature. Whereas the Egyptians indicated each higher unit by a new symbol, the Sumerians used the same symbol but indicated its value by its *position*. Thus, 1 followed by another 1 meant 61, and 5 followed by 6 followed by 3 (we shall write 5, 6, 3) meant $5 \times 60^2 + 6 \times 60 + 3 = 18363$. This position, or place-value, system did not differ essentially from our own system of writing numbers, in which the symbol 343 stands for $3 \times 10^2 + 4 \times 10 + 3$. Such a system had enormous advantages for computation, as we can readily see when we try to perform a multiplication in our own system and in a system with Roman numerals. The position system also removed many of the difficulties of fractional arithmetic, as does our own decimal system of writing fractions. This whole system seems to have developed as a direct result of the technique of administration, as is indicated in thousands of texts dating from the same period dealing with the delivery of cattle, grain, etc., and with arithmetical work based on these transactions.

Some ambiguities existed in this type of reckoning since the exact meaning of each symbol was not always clear from its position. Thus (5, 6, 3) might also

[6]See S. Gandz, *Quellen und Studien zur Geschichte der Mathematik*, Vol. I (1930), pp. 255–77.

mean $5 \times 60^1 + 6 \times 60^0 + 3 \times 60^{-1} = 306\ 1/20$; the exact interpretation had to be gathered from the context. Another uncertainty was introduced through the fact that a blank space sometimes meant zero, so that (11, 5) might stand for $11 \times 60^2 + 5 = 39605$. Eventually a special symbol for zero appeared, but not until the Persian era. The so-called "invention of the zero" was, therefore, a logical result of the introduction of the position system, but only after the technique of computation had reached a considerable perfection.

Both the sexagesimal system and the place-value system remained in the permanent possession of mankind. Our present division of the hour into 60 minutes and 3600 seconds dates back to the Sumerians, as does our division of the circle into 360 degrees, each degree into 60 minutes and each minute into 60 seconds. There is reason to believe that this choice of 60 rather than 10 as a unit occurred in an attempt to unify systems of measure, although the fact that 60 has many divisors may also have played a role. As to the place-value system, the permanent importance of which has been compared to that of the alphabet[7] (both inventions replaced a complex symbolism by a method easily understood by a large number of people), its history is still considerably obscure. It is reasonable to suppose that both Hindus and Greeks made its acquaintance on the caravan routes through Babylon; we also know that the Moslem scholars described it as an Indian invention. The Babylonian tradition, however, may have influenced all later acceptance of the position system.

5

The next group of cuneiform texts dates back to the first Babylonian Dynasty, when King Hammurabi reigned in Babylon (c. 1750 B.C.) and a Semitic population had subdued the original Sumerians. In these texts we find arithmetic evolved into a well-established algebra. Although the Egyptians of this period were only able to solve simple linear equations, the Babylonians of Hammurabi's days were in full possession of the technique of handling quadratic equations. They solved linear and quadratic equations in two variables, and even problems involving cubic and biquadratic equations. They formulated such problems only with specific numerical values for the coefficients, but their method leaves no doubt that they knew the general rule.

Following is an example taken from a tablet dating from this period, translated into modern language:[8]

> An area A, consisting of the sum of two squares, is 1000. The side of one square is 2/3 of the side of the other square, diminished by 10. What are the sides of the square?
> This leads to the equations $x^2 + y^2 = 1000$, $y = 2/3\ x - 10$, of which the solution can be found by solving the quadratic equation

[7]O. Neugebauer, "The History of Ancient Astronomy," *Journal of Near Eastern Studies*, Vol. 4 (1945), p. 12.

[8]K. Vogel, *Vorgriechische Mathematik*, Vol. II (Hanover and Paderborn, 1959), p. 50. The tablet is in Strasbourg's Bibliothèque Nationale et Universitaire.

$$\frac{13}{9}\,x^2 - \frac{40}{3}\,x - 900 = 0,$$

which has one positive solution, $x = 30$.

The actual solution in the cuneiform text confines itself—as in all Oriental problems—to the simple enumeration of the numerical steps that must be taken to solve the quadratic equation.

Square 10; this gives 100; subtract 100 from 1000; this gives 900, etc. The number 1000 is written (16, 40), 900 is (15, 0), etc.

The strong arithmetical-algebraic character of this Babylonian mathematics is also apparent from its geometry. As in Egypt, geometry developed from a foundation of practical problems dealing with mensuration, but the geometrical form of the problem was usually only a way of presenting an algebraic question. The previous example shows how a problem concerning the area of a square led to a nontrivial algebraic problem, and this example is no exception. The texts show that the Babylonian geometry of the Semitic period was in possession of formulas for the areas of simple rectilinear figures and for the volumes of simple solids, although the volume of a truncated pyramid had not yet been found. The so-called theorem of Pythagoras was known, not only for special cases, but in full generality, as a numerical relation between the sides of a right triangle. This led to the discovery of "Pythagorean triples" such as (3, 4, 5), (5, 12, 13), etc. The main characteristic of this geometry was, however, its support of the algebra. This is equally true of all later texts, especially those dating back to the third period of which we have a generous number of texts, that of the New Babylonian, Persian, and Seleucid eras (c. 600 B.C.–A.D. 300).

The texts of this later period are strongly influenced by the development of Babylonian astronomy, which in those days assumed really scientific traits, characterized by a careful analysis of the different ephemerides, also useful for astrological purposes. Mathematics became even more perfect in its computational technique; algebra tackled problems in equations which even now require considerable numerical skill. There exist computations dating from the Seleucid period which go to seventeen sexagesimal places. Such complicated numerical work was no longer related to problems of taxation or mensuration, but was stimulated by astronomical problems or by pure love of computation. "Oriental mathematics" was certainly not purely practical.

Much of this computational arithmetic was done with tables, which ranged from simple multiplication tables to lists of reciprocals and of square and cubic roots. One table gives a list of numbers of the form $n^3 + n^2$, which was used, it seems, to solve cubic equations such as $x^3 + x^2 = a$. There were some excellent approximations: $\sqrt{2}$ was indicated by $1\frac{5}{12}(\sqrt{2} = 1.4142\ldots, 1\frac{5}{12} = 1.4167)$,[9] and $1/\sqrt{2} = .7071$ by $17/24 = .7083$. Square roots seem to have been found by formulas like these:

---

[9]O. Neugebauer, *Exact Science in Antiquity*, Univ. of Pennsylvania Bicentennial Conf., Studies in Civilization (Philadelphia, 1941), pp. 13–29.

$$\sqrt{A} = \sqrt{a^2 + h} = a + h/2a = \frac{1}{2}\left(a + \frac{A}{a}\right).$$

In most Babylonian mathematics no better approximation for π has been found than the Biblical $\pi = 3$ (1 Kings vii: 23), the area of a circle being taken as $\frac{1}{12}$ of the square of its circumference, but approximations have been found that give a value $\pi = 3\frac{1}{8}$.[10]

The equation $x^3 + x^2 = a$ appears in a problem which calls for the solution of simultaneous equations $xyz + xy = 1 + 1/6$, $y = 2/3\,x$, $z = 12\,x$ which leads to $(12\,x)^3 + (12\,x)^2 = 252$, or $12\,x = 6$ (from the table).

There also exist cuneiform texts with problems in compound interest, such as the question of how long it would take for a certain sum of money to be doubled at 20 percent interest. This leads to the equation $(1\frac{1}{5})^x = 2$, which is solved by first remarking that $3 < x < 4$, and then by linear interpolation (in our way of writing):

$$4 - x = \frac{(1.2)^4 - 2}{(1.2)^4 - (1.2)^3},$$

leading to $x = 4$ years minus $(2, 33, 20)$ months.

One of the specific reasons for the development of algebra $c.$ 2000 B.C. seems to have been the use of the old Sumerian script by the new Semitic rulers, the Babylonians. The ancient script was, like the hieroglyphics, a collection of ideograms, each sign denoting a single concept. The Semites used them for the phonetic rendition of their own language and also took over some signs in their old meaning. These signs now expressed concepts, but were pronounced in a different way. Such ideograms were well fitted for an algebraic language, as are our present signs $+$, $-$, $:$, etc, which are really also ideograms. In the schools for administrators in Babylon this algebraic language became a part of the curriculum for many generations, and although the empire passed through the hands of many rulers—Kassites, Assyrians, Medes, Persians—the tradition remained.

The more intricate problems date back to later periods in the history of ancient civilization, notably to the Persian and Seleucid times. Babylon, in those days, was no longer an important political center, but remained for many centuries the cultural heart of a large empire, where Babylonians mixed with Persians, Greeks, Jews, Hindus, and many other peoples. In all the cuneiform texts there is a continuity of tradition which seems to point to a continuous local development. There is little doubt that this local development was also stimulated by the contact with other civilizations, and that this stimulation acted both ways. We know that Babylonian astronomy of this period influenced Greek astronomy and that Babylonian mathematics influenced computational arithmetic; it is reasonable to assume that through the medium of the Babylonian schools of scribes, Greek science and Hindu science met. The role of Persian and Seleucid Mesopotamia in the spread of ancient and antique astronomy and mathematics is still poorly known, but all available evidence shows that it must have been a considerable one. Medieval

---

[10]E. M. Bruins and M. Rutten, *Textes mathématiques de Suse* (Paris, 1961), p. 18.

ONE SIDE OF A CUNEIFORM TEXT NOW IN THE BRITISH MUSEUM.
Courtesy of the British Museum.

Arabic and Hindu science was based not only on the tradition of Alexandria but also on that of Babylon.

6

Nowhere in all ancient Oriental mathematics do we find any attempt at what we call a demonstration. No argumentation was presented, but only the prescription of certain rules: "Do such, do so." We are ignorant of the way in which the theorems were found: for instance, how did the Babylonians become acquainted with the theorem of Pythagoras? Several attempts exist to explain the way in which Egyptians and Babylonians obtained their results, but they are all of a hypothetical nature. To those who have been educated in Euclid's strict argumentation, the entire Oriental way of reasoning seems at first strange and highly unsatisfactory. But this strangeness wears off when we realize that most of the mathematics we teach our present-day engineers and technicians is still of the "do such, do so" type, without much attempt at rigorous demonstration. Algebra is still being taught in many high schools as a set of rules rather than as a science of deduction. Oriental mathematics, in this respect, never seems to have been emancipated from the millennial influence of the problems in technology and administration, for the use of which it had been invented.

7

The question of Greek, Chinese, and Babylonian influence determines profoundly the study of ancient Hindu mathematics. The native Indian and Chinese scholars of later days used to stress—and sometimes still do—the great antiquity of their mathematics, but there are no mathematical texts in existence which can be definitely dated to the pre-Christian era. The oldest extant Hindu texts are perhaps from the first centuries A.D.; the oldest Chinese texts date back to about the same period or a little earlier. We do know that the ancient Hindus used decimal systems of numeration without a place-value notation. Such a system was formed by the so-called Brāhmī numerals, which had special signs for each of the numbers 1, 2, 3, . . ., 9, 10; 20, 30, 40, . . ., 100; 200, 300, . . ., 1000; 2000, . . .; these symbols go back at least to the time of King Aśoka (c. 300 B.C.).

Then there exist the so-called *Śulvasūtras*, parts of which date back to 500 B.C. or earlier, and which contain mathematical rules which may be of ancient native origin. These rules are found among ritualistic prescriptions, some of which deal with the construction of altars. We find here recipes for the construction of squares and rectangles and expressions for the relation of the diagonal to the sides of the square and for the equivalence of circles and squares. There is some knowledge of the Pythagorean theorem in specific cases, and there are a few curious approximations in terms of unit fractions, such as (in our notation):

$$\sqrt{2} = 1 + \frac{1}{3} + \frac{1}{3.4} - \frac{1}{3.4.34} (= 1.4142156)$$

$$\pi = 4 \left(1 - \frac{1}{8} + \frac{1}{8.29} - \frac{1}{8.29.6} + \frac{1}{8.29.6.8}\right)^2 = 18(3 - 2\sqrt{2}).$$

The curious fact that these results of the *Śulvasūtras* do not occur in later Hindu works shows that we cannot yet speak of that continuity of tradition in Hindu mathematics which is so typical of its Egyptian and Babylonian counterparts; this continuity may actually be absent, India being as large as it is. There may have been different traditions relating to various schools. We know, for instance, that Jainism, which is as ancient as Buddhism (*c.* 500 B.C.) encouraged mathematical studies; in the sacred books of Jaina the value $\pi = \sqrt{10}$ is given.[11]

8

The study of ancient Chinese mathematics has long been handicapped by the lack of translated material, and most of us who are not Chinese scholars or natives have had to be satisfied with information obtainable in the books of Mikami (1913) and Needham (1959) or some special articles. At present several translated texts are available, among them a Russian as well as a German translation of the mathematical classic *Jiu zhang suan-shu*[12] (*Chiu chang suan shu*), or *Nine Chapters on the Mathematical Art*. Both this book and the *Zhou bei* (*Chou pei*) date in their present form from the period of the Han Dynasty (206 B.C.–A.D. 220), but may well contain much older material. The *Zhou bei* is only partly mathematical, and is of interest because it contains a discussion of the theorem of Pythagoras. However, the *Nine Chapters* is totally a mathematical book, already quite characteristic of the nature of ancient Chinese mathematics throughout the next thousand years and more.

Also very old are certain diagrams which can be found in books dating from the Han period, such as the *Yi-jing* (*I-ching*, or *Book of Changes*). One of these diagrams is the magic square (*lo-shu*):

4	9	2
3	5	7
8	1	6

with which many legends are connected.

Chinese numeration has always been decimal, and as early as the second millennium B.C. we find numbers expressed with nine symbols and with place value. The system was stabilized under the Han, or possibly earlier. The nine numerals were expressed by an arrangement of bamboo sticks, so that, for instance, ⊥ ⊤⊤ = ⊤⊤⊤⊤ meant 6729, and it was also written in this way. The elementary operations were carried out on counting boards, with blanks where we would put zero (a special symbol for zero appears only in the thirteenth century A.D., but may be older). In the computation of the calendar a kind of

---

[11]B. Datta, "The Jaina School of Mathematics," *Bull. Calcutta Math. Soc.*, Vol. 21 (1929), pp. 115–146.

[12]We spell Chinese words here in the Pinyin romanization, devised in 1956 and now more and more widely adopted. The most common romanization used before appears in parentheses. For the Pinyin romanization I have to thank Dr. Raymond Lum of the Harvard Library.

sexagesimal system was used, somewhat comparable to a combination of two connected cogwheels, one with 12, the other with 10 teeth, so that 60 became a higher unit, a "cycle" (the "cycle in Cathay" of Tennyson's poem).

The mathematics of the *Nine Chapters* consists mainly of a set of problems with general rules for their solution; they are of a computational arithmetic character and lead to algebraic equations with numerical coefficients. Square and cubic roots are drawn: e.g., $751\frac{1}{2}$ is found as the square root of $564752\frac{1}{4}$. For circle measurements, π is taken as 3. A series of problems leads to systems of linear equations, e.g., to the system

$$3x + 2y + z = 39$$
$$2x + 3y + z = 34$$
$$x + 2y + 3z = 26,$$

which is written in the form of the matrix of their coefficients. The solution is performed by what we now would call matrix transformations. In these matrices we find negative numbers, which appear here for the first time in history.

Chinese mathematics is in the exceptional position that its tradition has remained practically unbroken until recent years, so that we can study its position in the community somewhat better than that of Egyptian and Babylonian mathematics, which belonged to vanished civilizations. We know, for instance, that candidates for examination had to display a precisely circumscribed knowledge of the classics, and that this examination was based mainly on the ability to cite texts correctly from memory. The traditional lore was thus transmitted from generation to generation with painful conscientiousness. In such a stagnant cultural atmosphere new discoveries became extraordinary exceptions, and this again guaranteed the invariability of the mathematical tradition. Such a tradition might be transmitted over millennia, only occasionally shaken by great historical catastrophes. In India a similar condition existed; here we even have examples of mathematical texts written in metric stanzas to facilitate memorization. There is no particular reason to believe that the ancient Egyptian and Babylonian practice may have been much different from the Indian and Chinese one. The emergence of an entirely new civilization was necessary to interrupt the relative ossification of mathematics. The different outlook on life characteristic of Greek civilization at last brought mathematics up to the standards of a new type of science.

Literature

The Rhind Mathematical Papyrus, T. E. Peet, ed. London, 1923.
The Rhind Mathematical Papyrus, A. B. Chase *et al.*, eds. 2 vols., Ohio, 1927–29. (Includes an extensive bibliography of Egyptian and Babylonian mathematics. See also the bibliography, mostly on ancient astronomy, in Neugebauer, *Exact Science in Antiquity*, Philadelphia, 1941, p. 18.)

Mathematischer Papyrus des staatlichen Museums der schönen Künste in Moskau, W. W. Struve and B. A. Turajeff, eds. Berlin, 1930.

Gillings, R. J. *Mathematics in the Time of the Pharaohs*. Cambridge, Mass., 1972 (Dover reprint, 1982). (See *HM*, Vol. 4 (1977), pp. 445–52, article by M. Bruckheimer and Y. Salomon.)

Neugebauer, O. *Vorlesungen über Geschichte der antiken mathematischen Wissenschaften, I: Vorgriechische Mathematik*. Berlin, 1934.

——. *Mathematische Keilschrift-Texte*. 3 vols., Berlin, 1935–37.

——. *The Exact Sciences in Antiquity*. Princeton, 1952; 2nd ed., 1957; Dover reprint, 1969.

—— and Sachs, A. *Mathematical Cuneiform Texts*. New Haven, 1945.

Thureau-Dangin, F. "Sketch of a History of the Sexagesimal System." *Osiris*, Vol. 7 (1939), pp. 95–141.

——. *Textes mathématiques babyloniens*. Leiden, 1938.

There is some difference in the interpretation of Babylonian mathematics by the two preceding authors. An opinion is expressed in:

Gandz, S. "Conflicting Interpretations of Babylonian Mathematics," *Isis*, Vol. 31 (1940), pp. 405–25.

See also:

Bruins, E. M., and Rutten, M. *Textes mathématiques de Suse*. Paris, 1961.
Vogel, K. *Vorgriechische Mathematik*. 2 vols. Hanover-Paderborn, 1958–59.

An older survey of pre-Greek mathematics is given in:

Archibald, R. C. "Mathematics Before the Greeks." *Science*, Vol. 71 (1930), pp. 109–21, 342. See also *ibid.*, Vol. 72 (1930), p. 36.
Smith, D. E. "Algebra of 4000 Years Ago." *Scripta math.*, Vol. 4 (1936), pp. 111–25.

On Indian mathematics, see the volumes of the *Bulletin of the Calcutta Mathematical Society* and:

Datta, B., and Singh, A. N. *History of Hindu Mathematics*. 2 vols., Lahore, 1935–38. [Reviewed by O. Neugebauer in *Quellen und Studien*, Vol. 3B (1936), pp. 263–71.]
Gurjar, L. V. *Ancient Indian Mathematics and Vedha*. Poona, 1947. [See also *Math. Rev.*, Vol. 9 (1948), p. 73.]
Kaye, G. R. "Indian Mathematics." *Isis*, Vol. 2 (1919), pp. 326–56.
Seidenberg, A. "The Ritual Origin of Geometry." *AHES*, Vol. 1 (1962), pp. 488–527.
Müller, C. "Die Mathematik der Śulvasūtra." *Abh. math. Seminar Univ. Hamburg*, Vol. 7 (1929), pp. 173–204.

On Chinese-Japanese mathematics see:

Mikami, Y. *The Development of Mathematics in China and Japan*. Leipzig, 1913.
Berezkina, E. I. "The Ancient Chinese Treatise 'Mathematics in Nine Chapters.' " *Istor.-mat. Issled.*, Vol. 10 (1957), pp. 423–584 (in Russian).

German edition:

Chiu Chang Suan Shu, Neun Bücher arithmetischer Technik, übers. und erläutert von K. Vogel. Ostwalds Klassiker der exakten Wissenschaften, Neue Folge 4. Braunschweig, 1908.

Needham, J. *Science and Civilization in China.* Cambridge, 1959. (See Vol. III, pp. 1–168.) This is, at present, our main source of information.

Haudricourt, A., and Needham, J. "La science chinoise antique." In *Histoire générale des Sciences* (Paris, 1957), Vol. 1, pp. 184–201.

Libbrecht, U. *Chinese Mathematics in the Thirteenth Century: The Shu-Shu Chiu-chang of Ch'in Chiu-shao.* Cambridge, Mass., 1973.

Lam, L. Y. *A Critical Study of the Yang Hui Suan Fa. A Thirteenth Century Chinese Mathematical Treatise.* Singapore, 1977.

The I Ching or *Book of Changes.* Trans. by R. Wilhelm, New York, 1950; and by J. Legge, London, 1899 (Dover reprint, 1963).

Gillon, B. S. "Introduction, Translation and Discussion of Chao Chun-ch'ing's Notes to the Diagram of Short Legs and Long Legs and of Squares and Circles." *HM*, Vol. 4 (1977), pp. 253–93.

Struik, D. J. "On Ancient Chinese Mathematics." *Mathematics Teacher*, Vol. 56 (1963), pp. 424–31; also *Euclides*, Vol. 40 (1964), pp. 65–79.

On the nature of Oriental society see the following and the Literature given in Chapter IV.

Wittfogel, K. A. "Die Theorie der orientalischen Gesellschaft." *Zeitschrift für Sozialforschung*, Vol. 7 (1938), pp. 90–122. Also "Le mode de production asiatique." *La Pensée*, Vol. 114 (1964), pp. 3–73.

Needham, J. "Science and Society in East and West." *Science and Society*, Vol. 28 (1964), pp. 385–408.

See further on Oriental mathematics:

van der Waerden, B. L. *Science Awakening.* 2nd ed., New York, 1961. (Trans. from the Dutch: Groningen, 1950.)

Vol. XV, Supplement I of *DSB* (New York, 1978) has on pp. 531–818 "Topical Essays" on "Mathematical Astronomy in India" (D. Pingree); "Man and Nature in Mesopotamian Civilization" (A. L. Oppenheim); "Mathematics and Astronomy in Mesopotamia" (B. L. van der Waerden); "The Mathematics of Ancient Egypt" (R. J. Gillings); "Egyptian Astronomy, Astrology and Calendrical Reckoning" (R. A. Pinker); "Japanese Scientific Thought" (S. Nakayama); and "Maya Numeration, Computation and Calendrical Astronomy" (F. C. Lounsbury).

van der Waerden, B. L. "On Pre-Babylonian Mathematics" (I, II). *AHES*, Vol. 23 (1980), pp. 1–26, 27–46. (On a possible prehistoric source of Chinese, Babylonian, and Egyptian mathematics, see Seidenberg, *AHES*, Vol. 18 (1978), pp. 301–42.)

CHAPTER III

Greece

1

Enormous economic and political changes occurred in and around the Mediterranean basin during the last centuries of the second millennium. In a turbulent atmosphere of migrations and wars the Bronze Age was replaced by what has been called our age, the Age of Iron. Few details are known about this period of revolutions, but we find that toward its end, perhaps *c.* 900 B.C., the Minoan, Mycenaean, and Hittite worlds had disappeared, the power of Egypt and Babylonia had been greatly reduced, and new peoples had come into a historical setting. The most outstanding of these new peoples were the Hebrews, the Assyrians, the Phoenicians, and the Greeks. The replacement of bronze by iron brought not only a change in warfare but, by cheapening the tools of production, increased the social surplus, stimulated trade, and allowed larger participation of the common people in matters of economy and public interest. This was reflected in two great innovations: the replacement of the clumsy script of the ancient Orient by the easy-to-learn alphabet, and the introduction of coined money, which helped to stimulate trade. The time had come when the religious and scientific cults could no longer be the exclusive province of an Oriental officialdom.

The activities of the "sea-raiders," as some of the migrating peoples are styled in Egyptian texts, were originally accompanied by great cultural losses. Minoan and Mycenaean civilization disappeared; Egyptian art declined; Babylonian and Egyptian science stagnated for centuries. No mathematical texts have come to us from this transition period. When stable relations were again established, the ancient Orient recovered mainly along traditional lines, but the stage was set for an entirely new type of civilization, the civilization of Greece, in which Homer's poems preserved about the only verbal memory of the Mycenaean past.

The towns which arose along the coast of Asia Minor and on the Greek mainland were no longer administration centers of an Oriental despotism. They were trading towns in which the old-time feudal landlords had to fight a losing battle with an independent, politically conscious merchant class. During the seventh and sixth centuries B.C., this merchant class won ascendancy and had to fight its own battles with the small traders and artisans, the *demos*. The result was the rise of the Greek *polis*, the self-governing city-state, a new social experiment entirely different from the early city-states of Sumer and other Oriental countries. The most important of these city-states developed in Ionia on the Anatolian coast. Their growing trade connected them with the shores of the whole Mediterranean, with Mesopotamia, Egypt, Scythia, and even with countries beyond. Miletus for a long time took a leading place. Cities on other shores also gained in wealth and importance: on the mainland of Greece, first Corinth, later Athens; on the Italian coast, Croton and Tarentum; in Sicily, Syracuse.

This new social order created a new type of man. The merchant trader had never enjoyed so much independence, but he knew that this independence was a result of a constant and bitter struggle. The static outlook of the Orient could never be his. He lived in a period of geographical discoveries comparable only to those of sixteenth-century Western Europe; he recognized no absolute monarch and no power supposedly vested in a static deity. Moreover, he could enjoy a certain amount of leisure, the result of wealth and, partly at any rate, of slave labor. He could philosophize about his world. The absence of any well-established religion might well have led many inhabitants of these coastal towns into some form of mysticism, but also stimulated the opposite, the growth of rationalism and the scientific outlook.

2

Modern mathematics was born in this atmosphere of Ionian rationalism—the mathematics which not only asked the Oriental question "How?" but also the modern scientific question "Why?" The traditional father of Greek mathematics is the merchant Thales of Miletus, who visited Babylon and Egypt in the first half of the sixth century B.C. And even if his whole figure is legendary, it stands for something eminently real. It symbolizes the circumstances under which the foundations, not only of modern mathematics, but also of modern science and philosophy, were established.

The early Greek study of mathematics had one main goal: the understanding of man's place in the universe according to a rational scheme. Mathematics helped to find order in chaos, to arrange ideas in logical chains, to find fundamental principles. It was the most rational of all sciences, and although there is little doubt that the Greek merchants became acquainted with Oriental mathematics along their trade routes, they soon discovered that the Orientals had left most of the rationalization undone. Why did the isosceles triangle have two equal angles? Why was the area of a triangle equal to half that of a rectangle of equal base and altitude? These questions came naturally to men who asked similar questions concerning cosmology, biology, and physics.

It is unfortunate that there are no primary sources which can give us a picture of the early development of Greek mathematics. The existing codices are from Christian and Islamic times, and they are only sparingly supplemented by Egyptian papyrus notes of a somewhat earlier date. Classical scholarship, however, has enabled us to restore the remaining texts, which date back to the fourth century B.C. and later, and thus we possess reliable editions of Euclid, Archimedes, Apollonius, and other great mathematicians of antiquity. But these texts represent an already fully developed mathematical science, in which historical development is hard to trace even with the aid of later commentaries. For information about the formative years of Greek mathematics we must rely entirely on small fragments transmitted by later authors and on scattered remarks by philosophers and other not strictly mathematical authors. Highly ingenious and patient text criticism has been able to elucidate many obscure points in this early history, and it is due to this work (carried on by investigators such as Paul Tannery, T. L. Heath, H. G. Zeuthen, and E. Frank, and which is still being carried on) that we are able to present a somewhat consistent, if somewhat hypothetical, picture of Greek mathematics in its formative years.

3

In the sixth century B.C. a new and vast Oriental power arose on the ruins of the Assyrian Empire: the Persia of the Achaemenids. It conquered the Anatolian towns, but the social structure on the mainland of Greece was already too well established to suffer defeat. The Persian invasion was repelled in the historic battles of Marathon, Salamis, and Plataea. The main result of the Greek victory was the expansion and hegemony of Athens. Here, under Pericles, in the second half of the fifth century B.C., democratic elements became increasingly influential. They were the driving force behind the economic and military expansion, and by 430 B.C. they made Athens not only the leader of a Greek Empire, but also the center of a new and amazing civilization—the Golden Age of Greece.

Within the framework of the social and political struggles, philosophers and teachers presented their theories and with them the new mathematics. For the first time in history, a group of critical men, the "sophists," less hampered by tradition than any previous group of learned persons, approached problems of a mathematical nature as part of a philosophical investigation of the natural and moral world. Thus developed a mathematics investigated in the spirit of understanding rather than of utility. As this mental attitude enabled the sophists to reach toward the foundations of exact thinking itself, it would be highly instructive to follow their discussions. Unfortunately, only one complete mathematical fragment of this period is extant; it is written by the Ionian philosopher, Hippocrates of Chios.[1] This fragment represents a high degree of perfection in mathematical reasoning and deals, typically enough, with a curiously "impractical" but theoretically valuable subject, the so-called *lunulae*—the little moons or crescents bounded by two or more circular arcs.

The subject—to find certain areas bounded by two circular arcs which can be expressed rationally in terms of the diameters—has a direct bearing on the problem of the quadrature of the circle, a central problem in Greek mathematics. In the analysis of his problem[2] Hippocrates showed that the mathematicians of the Golden Age of Greece had an ordered system of plane geometry, in which the

---

[1]Not to be confused with the physician Hippocrates of Cos, roughly contemporary.

[2]For a modern analysis see E. Landau, "Über quadrirbare Kreisbogenzweiecke," *Berichte Berliner Math. Ges.*, Vol. 2 (1903), pp. 1–6. Also: T. Dantzig, *The Bequest of the Greeks* (New York, 1955), Ch. 10; and *DSB*, Vol. 6 (1972), pp. 411–16.

principle of logical deduction from one statement to another (*apagoge*) had been fully accepted. A beginning of axiomatics had been made, as is indicated by the name of the book supposedly written by Hippocrates, the *Elements* (*stoicheia*), the title of all Greek axiomatic treatises including that of Euclid. Hippocrates investigated the areas of plane figures bounded by straight lines as well as circular arcs. The areas of similar circular segments, he taught, are to each other as the squares of their chords. Pythagoras' theorem was known to him and so was the corresponding inequality for nonrectangular triangles. The whole treatise is already in what might be called the Euclidean tradition, but it is older than Euclid by more than a century.

The problem of the quadrature of the circle is one of the "three famous mathematical problems of antiquity," which in this period began to be a subject of study. These problems were as follows:

1. The trisection of the angle; that is, the problem of dividing a given angle into three equal parts.

2. The duplication of the cube; that is, to find the side of a cube of which the volume is twice that of a given cube (the so-called *Delian* problem).

3. The quadrature of the circle; that is, to find the square of an area equal to that of a given circle.

The importance of these problems lies in the fact that they cannot be geometrically solved by the construction of a finite number of straight lines and circles except by approximation, and therefore they served as a means of penetration into new fields of mathematics. The first two problems were often reduced to the search for two line segments x and y such that $a:x = x:y = y:b$, where x and b are given line segments. This problem is an extension of the search for an x for which $a:x = x:b$, the geometric proportional, but the search for the double geometric proportional cannot be solved with compass and ruler alone. This again led to the discovery of the conic sections, and some cubic and quartic curves, and to one transcendental curve, the *quadratrix*. The anecdotic forms in which the problems have occasionally been transmitted (Delphic oracles, etc.) should not prejudice us against their fundamental importance. It occurs not infrequently that a fundamental problem is presented in the form of an anecdote or a puzzle—Newton's apple, Cardan's broken promise, or Kepler's wine barrels. Mathematicians of different periods, including our own, have shown the connection between these Greek problems and the modern theory of equations, involving considerations concerning domains of rationality, algebraic numbers, and group theory.[3]

4

Probably aside from the group of sophists, who can to some degree be thought of as connected with the democratic movement, there existed another group of mathematically inclined philosophers related to the aristocratic factions. They

---

[3]See, e.g., F. Klein, *Vorträge über ausgewählte Fragen der Elementargeometrie* (Leipzig, 1895); and F. Enriques, *Fragen der Elementarmathematik* (Leipzig, 1907), Vol. II.

were called Pythagoreans, after a rather mythical founder of the school, Pythagoras, supposedly a mystic, a scientist, and an aristocratic statesman. While most sophists emphasized the reality of change—in particular, the Atomists, followers of Leucippus and Democritus—the Pythagoreans stressed the study of the unchangeable elements in nature and society. In their search for the eternal laws of the universe they studied geometry, arithmetic, astronomy, and music (later called the *quadrivium*). Their most outstanding leader was Archytas of Tarentum, who lived at about 400 B.C., and to whose school, if we follow the hypothesis of E. Frank, much of the Pythagorean brand of mathematics may be ascribed. Its arithmetic was a highly speculative science which had little in common with the contemporary Babylonian computational technique. Numbers (i.e., integers, called *arithmoi*) were divided into classes: odd, even, even-times-even, odd-times-odd, prime and composite, perfect, friendly, triangular, square, pentagonal, etc. Some of the most interesting results concern the "triangular numbers," which represent a link between geometry and arithmetic:

. 1   . .   3,   . . .   6, ...   10,   etc.

Our name "square numbers" had its origin in Pythagorean speculations:

. 1,   . .   4,   . . .   9,   etc.

The figures themselves are much older, since some of them appear on neolithic pottery. The Pythagoreans investigated their properties, adding their brand of number mysticism and placing them in the center of a cosmic philosophy which tried to reduce all fundamental relations to number relations ("everything is number"). A point was "unity in position."[4] Of particular importance was the ratio of numbers (*logos*, Lat. *ratio*). Equality of ratios formed a proportion. They discriminated between an arithmetical $(2b = a + c)$, geometrical $(b^2 = ac)$, and harmonical $\left(\dfrac{2}{b} = \dfrac{1}{a} + \dfrac{1}{c}\right)$ proportion that they interpreted philosophically and socially.

The Pythagoreans knew some properties of regular polygons and regular bodies. They showed how the plane can be filled by means of patterns of regular triangles, squares, or regular hexagons, and space by cubes, to which Aristotle later tried to add the wrong notion that space can be filled by regular tetrahedra.[5] The Pythagoreans may have also known the regular octahedron and dodecahe-

[4]On the arithmetic of the Pythagoreans see B. L. van der Waerden, *Math. Annalen*, Vol. 120 (1948), pp. 127–53, 676–700.
[5]D. J. Struik, *Nieuw Arch. v. Wiskunde*, Vol. 15 (1925), pp. 121–37; M. Senechal, "Which Tetrahedra Fill Space?" *Mathematics Magazine*, Vol. 54 (1981), pp. 227–43.

dron—the latter figure because pyrite, found in Italy, crystallizes in dodecahedra, and models of figures such as ornaments or magical symbols date back to Etruscan times. They date back to Celtic peoples of Central Europe during the beginnings of the Iron Age, c. 900 B.C., and later (pyrite is a source of iron).[6]

As to Pythagoras' theorem, the Pythagoreans ascribed its discovery to their master, who was supposed to have sacrificed a hundred oxen to the gods as a token of gratitude. We have seen that the theorem was already known in Hammurabi's Babylon, but the first general proof may very well have been obtained in the Pythagorean school. Where the Babylonians saw it primarily as an achievement in mensuration, the Pythagoreans conceived it as an abstract geometrical theorem.

The most important discovery ascribed to the Pythagoreans was the discovery of "the irrational" by means of incommensurable line segments. This discovery may have been the result of their interest in the geometric mean $a:b = b:c$, which served as a symbol of aristocracy. What was the geometric mean of 1 and 2, two sacred symbols? This led to the study of the ratio of the side and the diagonal of the square, and it was found that this ratio could not be expressed by "numbers"—that is, by what we now call rational numbers (integers or fractions), the only numbers recognized as such. This can be seen as follows, according to Aristotle.[7]

> Suppose this ratio to be $p:q$, in which we can always take p and q as relative primes. Then $p^2 = 2q^2$, hence p^2, and therefore p, is even, say $p = 2r$. Then q must be odd; but since $q^2 = 2r^2$, it must also be even. This contradiction was not solved, as in the Orient or in Renaissance Europe, by an extension of the conception of number, but by rejecting the theory of numbers for such cases and looking for a synthesis in geometry.

This discovery, which upset the easy harmony between arithmetic and geometry, was probably made in the last decades of the fifth century B.C. It came on top of another difficulty, which had emerged from the arguments concerning the reality of change, arguments which have kept philosophers busy from then until the present. This difficulty has been ascribed to Zeno of Elea (c. 450 B.C.), a pupil of Parmenides, a conservative philosopher who taught that reason only recognizes absolute being, and that change is only apparent. This received mathematical significance when infinite processes had to be studied in questions such as the determination of the volume of a pyramid. Here Zeno's paradoxes conflicted with some ancient and intuitive conceptions concerning the infinitely small and the infinitely large. It was always believed that the sum of an infinite number of quantities can be made as large as we like, even if each quantity is extremely small ($\infty \times \epsilon = \infty$), and also that the sum of a finite or infinite number of quantities of dimension zero is zero ($n \times 0 = 0$, $\infty \times 0 = 0$). Zeno's criticism challenged these conceptions, and his four paradoxes created a stir, the ripples of

---

[6] F. Lindemann, *Sitzungsberichte bayr. Akad. Wiss., München*, Vol. 26 (1897), pp. 625–758. Also, *ibid.* (1934), pp. 265–75.

[7] Aristotle, in his *Analytica Priora* (I, 23), only hints at this proof as an example of *reductio ad absurdum*. For the proof see T. L. Heath, *The Thirteen Books of Euclid's Elements* (Dover reprint, 1956), Vol. III, p. 2.

which can be observed today. They have been preserved by Aristotle and are known as the *Achilles*, the *Arrow*, the *Dichotomy*, and the *Stadium*. They are phrased so as to stress contradictions in the conception of motion and of time; no satisfactory attempt is made to solve the contradictions.

The gist of the reasoning will be clear from the *Achilles* and the *Dichotomy*, which we explain in our own words as follows:

Achilles. Achilles and a tortoise move in the same direction on a straight line. Achilles is much faster than the tortoise, but in order to reach the tortoise he must first pass the point P from which the tortoise started. If he comes to P, the tortoise has advanced to point P_1. Achilles cannot reach the tortoise until he has passed P_1, but the tortoise has advanced to a new point P_2. If Achilles is at P_2, the tortoise has reached a new point P_3, etc. Hence Achilles can never reach the tortoise.

Dichotomy. Suppose I like to go from A to B along a line. In order to reach B, I must first traverse half the distance AB_1 of AB, and in order to reach B_1 I must first reach B_2 halfway between A and B_1. This goes on indefinitely so that the motion can never even *begin*.

Zeno's arguments showed that a finite segment can be broken up into an infinite number of small segments, each of finite length. They also showed that there is a difficulty in explaining what we mean by saying that a line is "composed of" points. It is very likely that Zeno himself had no idea of the mathematical implications of his arguments. Problems leading to his paradoxes have regularly appeared in the course of philosophical and theological discussions; we recognize them as problems concerning the relation of the "potentially" and the "actually" infinite. Paul Tannery, however, believed that Zeno's arguments were particularly directed against the Pythagorean idea of space as the sum of points ("the point is unity in position").[8] Whatever the truth may be, Zeno's reasoning certainly influenced mathematical thought for many generations. His paradoxes may be compared to those used by Bishop Berkeley in 1734, when he showed the logical absurdities to which poor formulation of the principles of the calculus may lead, but without offering a better foundation.

Zeno's arguments began to worry the mathematicians even more after the irrational had been discovered. Was mathematics as an exact science possible? Tannery[9] has suggested that we may speak of "a veritable logical scandal"—of a "crisis" in Greek mathematics.[10] If this is the case, then this crisis originated in the later period of the Peloponnesian War, ending with the fall of Athens (404 B.C.). We may then detect a connection between the crisis in mathematics and that of the social system, since the fall of Athens spelled the doom of the empire of a slave-owning democracy and introduced a new period of aristocratic supremacy—a crisis which was solved in the spirit of the new period.

---

[8] P. Tannery, *La géométrie grecque* (Paris, 1887), pp. 217–61. Another opinion is given in B. L. van der Waerden, *Math. Annalen*, Vol. 117 (1940), pp. 141–61.

[9] *Ibid.*, p. 98. Tannery, at this place, deals only with the breakdown of the ancient theory of proportions as a result of the discovery of incommensurable line segments.

[10] See on this H. Freudenthal, "Y avait-il une crise des fondements des mathématiques dans l'Antiquité?" *Bull. soc. math. Belgique*, Vol. 18 (1966), pp. 43–55.

5

Typical of this new period in Greek history was the increasing wealth of certain sections of the ruling classes combined with equally increased misery and insecurity of the poor. The ruling classes based their material existence more and more upon slavery, which allowed them leisure to cultivate arts and sciences, but made them also more and more averse to all manual work. A gentleman of leisure looked down upon the work of slaves and craftsmen and sought relief from worry in the study of philosophy and of personal ethics. Plato and Aristotle expressed this attitude; and it is in Plato's *Republic* (written, perhaps, *c.* 360 B.C.) that we find the clearest expression of the ideals of the slave-owning ruling class. The "guards" of Plato's *Republic* must study the *quadrivium*, consisting of arithmetic, geometry, astronomy, and music, in order to understand the laws of the universe.[11] Such an intellectual atmosphere was conducive (in its earlier period, at any rate) to a discussion of the foundations of mathematics and to speculative cosmogony. At least three great mathematicians of this period were connected with Plato's Academy, namely, Archytas, Theaetetus (d. 369 B.C.), and Eudoxus (*c.* 408–355 B.C.). Theaetetus has been credited with the theory of irrationals as it appears in the tenth book of Euclid's *Elements*. Eudoxus' name is connected with the theory of proportions which Euclid gave in his fifth book, and also with the so-called "exhaustion method," which allowed a rigorous treatment of area and volume computations. This means that it was Eudoxus who solved the "crisis" in Greek mathematics, and whose rigorous formulations helped to determine the course of Greek axiomatics and, to a considerable extent, of Greek mathematics as a whole.

Eudoxus' theory of proportions did away with the arithmetical theory of the Pythagoreans, which applied to commensurable quantities only. It was a purely geometrical theory, which, in strictly axiomatic form, made all reference to incommensurable or commensurable magnitudes superfluous.

Typical is Definition 5, Book V, of Euclid's *Elements*:

> Magnitudes are said to be in the same ratio, the first to the second and third to the fourth, when, if any equimultiples whatever be taken of the first and third, and any equimultiples whatever of the second and fourth, the former equimultiples alike exceed, are alike equal to, or alike fall short of, the latter equimultiples taken in corresponding order.[12]

This means, in our notation, that $a:b = c:d$, if $ma > nb$ implies $mc > nd$, $ma = nb$ implies $mc = nd$, and $ma < nb$ implies $mc < nd$, where m and n are integers. The possibility of such a definition first had to be established by the so-called Axiom of Archimedes, which in Euclid's *Elements* precedes the previously given definition in the form of Definition 4:

---

[11]The gate to Plato's Academy carried, according to a late source, the motto: "Let no one ignorant of geometry enter."

[12]T. L. Heath, *The Thirteen Books of Euclid's Elements* (Cambridge, 1912), Vol. 2, p. 114. (Dover reprint, 1956.)

PORTION OF A PAGE OF THE FIRST EDITION OF EUCLID'S "ELEMENTS," 1482.

> Magnitudes are said to have a ratio to one another which are capable, when multiplied, of exceeding one another.[13]

This definition could better be called the Axiom of Eudoxus. The present theory of irrational numbers, developed by Dedekind and Weierstrass, follows Eudoxus' mode of thought almost literally, but by using modern arithmetical methods has opened far wider perspectives.

The "exhaustion method" (the term "exhaust" appears first in Grégoire de Saint-Vincent, 1647) was the Platonic school's answer to Zeno. It avoided the pitfalls of the infinitesimals by simply discarding them, by reducing problems which might lead to infinitesimals to problems involving formal logic only. For instance, when it was required to prove that the volume V of a tetrahedron is equal to one-third the volume P of a prism of equal base and altitude, the proof consisted in showing that both the assumptions $V > \frac{1}{3}P$ and $V < \frac{1}{3}P$ lead to absurdities. For this purpose an axiom was introduced equivalent to the Axiom of Archimedes (or Eudoxus), and which Archimedes formulated as follows: that of two unequal magnitudes "the greater exceeds the less by such magnitude as, when added to itself, can be made to exceed any assigned magnitude among those which are comparable with it and with one another."[14] Here the "adding to itself" can be repeated any number of times. In our case of the tetrahedron, the reasoning then was: that the hypothesis $V = A$, $A > \frac{1}{3}P$ is shown to be absurd by enclosing the tetrahedron in a circumscribed step-pyramid of n prisms, each of height h/n, where h is the altitude of the tetrahedron, and then to show that by the choice of a sufficiently large n the volume of the step-pyramid could be made $<A$. Since this volume is certainly $>V$, we reach an absurdity. In a similar way we show the absurdity of $V < \frac{1}{3}P$ by an inscribed step-pyramid. Euclid proves in this way several propositions, e.g., the theorem that two circles are to each other as the squares of the diameters.

This indirect method, which became the standard Greek and Renaissance mode of strict proof in area and volume computation, was quite rigorous, and can easily be translated into a proof satisfying the requirements of modern analysis. It had the great disadvantage that the result, in order to be proved, must be known in advance, so that the mathematician finds it first by another less rigorous and more tentative method.

There are clear indications that another such method was actually used. We possess a letter from Archimedes to Eratosthenes (c. 250 B.C.) which was not discovered until 1906, in which Archimedes described a nonrigorous but fertile way of finding results. This letter is known as the "Method." It has been suggested, notably by S. Luria, that it represented a school of mathematical reasoning competing with the school of Eudoxus, also dating back to the period of the "crisis" and associated with the name of Democritus, the founder of the atom theory. In Democritus' school, according to the theory of Luria, the notion of the

---

[13]*Ibid.*, p. 114.

[14]Archimedes, *On the Sphere and Cylinder*, Book I, Assumption 5. See T. L. Heath, *The Works of Archimedes* (Cambridge, 1897), p. 4.

"geometrical atom" was introduced. A line segment, an area, or a volume was supposed to be built up of a large, but finite, number of indivisible "atoms." The computation of a volume was the summation of the volumes of all the "atoms" of which this body consists. This theory sounds perhaps absurd, until we realize that several mathematicians in the period before Newton, notably Kepler, used essentially the same conceptions, taking the circumference of a circle as composed of a very large number of tiny line segments. There is no evidence that antiquity ever developed a rigorous method on this foundation, but our modern limit conceptions have made it possible to build this "atom" theory into a theory as rigorous as the "exhaustion method." Even today we use this conception of the "atoms" quite regularly in setting up a mathematical problem in the theory of elasticity, in physics or in chemistry, reserving the rigorous "limit" theory for the professional mathematician.[15]

The advantage of the "atom method" over the "exhaustion method" was that it facilitated the finding of new results. Antiquity had thus the choice between a rigorous but relatively sterile, and a loosely founded but far more fertile method. It is instructive that in practically all the classical texts the first method was used. This again may be connected with the fact that mathematics had become a hobby of a leisure class which was based on slavery, indifferent to invention, and interested in contemplation. It may also be a reflection of the victory of Platonic idealism over Democritean materialism in the realm of mathematical philosophy.

6

In 334 B.C. Alexander the Great began his conquest of Persia. When he died at Babylon in 323, the whole Near East had fallen to the Greeks. Alexander's conquests were divided among his generals and eventually three empires emerged: Egypt under the Ptolemies; Mesopotamia and Syria under the Seleucids; and Macedonia under Antigonus and his successors. Even the Indus valley had Greek princes. The period of Hellenism had begun.

The immediate consequence of Alexander's campaign was the acceleration of the advance of Greek civilization over large sections of the Oriental world. Egypt and Mesopotamia and a part of India were Hellenized. The Greeks flooded the Near East as traders, merchants, physicians, adventurers, travelers, and mercenaries. The cities—many of them newly founded and recognizable by their Hellenistic names—were under Greek military and administrative control, and had a mixed population of Greeks and Orientals. But Hellenism was essentially an urban civilization. The countryside remained native and continued its existence in the traditional way. In the cities the ancient Oriental culture met with the imported civilization of Greece and partly mixed with it, although there always remained a deep separation between the two worlds. The Hellenistic monarchs

---

[15]"Thus, so far as first differentials are concerned, a small part of a curve near a point may be considered straight and a part of a surface plane; during a short time a particle may be considered as moving with constant velocity and any physical process as occurring at a constant rate." [H. B. Phillips, *Differential Equations* (New York, 1922), p. 7.]

adopted Oriental manners and had to deal with Oriental problems of administration, but they stimulated Greek arts, letters, and sciences.

Greek mathematics, thus transplanted to new surroundings, kept many of its traditional aspects, but experienced also the influence of the problems in administration and astronomy which the Orient had to solve. This close contact of Greek science with the Orient was extremely fertile, especially during the first centuries. Practically all the really productive work which we call "Greek mathematics" was produced in the relatively short interval from 350 to 200 B.C., from Eudoxus to Apollonius, and even Eudoxus' achievements are known to us only through their interpretation by Euclid and Archimedes. It is also remarkable that the greatest flowering of this Hellenistic mathematics occurred in Egypt under the Ptolemies and not in Mesopotamia, despite the more advanced status of native mathematics in Babylonia.

The reason for this development may be found in the fact that Egypt was now in a central position in the Mediterranean world. Alexandria, the new capital, was built on the seacoast, and became the intellectual and economic center of the Hellenistic world. But Babylon lingered on only as a remote center of caravan roads, and eventually disappeared, to be replaced by Ctesiphon-Seleucia, the new capital of the Seleucids. As far as we know, no great Greek mathematicians were ever connected with Babylon. Antioch and Pergamum, also cities of the Seleucid Empire but closer to the Mediterranean, had important Greek schools. The development of native Babylonian astronomy and mathematics reached its height under the Seleucids, and Greek astronomy received an impetus, the importance of which is only now beginning to be better understood. Besides Alexandria there were other centers of mathematical learning, especially Athens and Syracuse. Athens became an educational center, while Syracuse produced Archimedes, the greatest of Greek mathematicians.

7

In this period the professional scientist appeared, a man who devoted his life to the pursuit of knowledge and received a salary for doing so. Some of the most outstanding representatives of this group lived in Alexandria, where the Ptolemies built a great center of learning in the so-called Museum with its famous Library. Here the Greek heritage in science and literature was preserved and developed. The success of this enterprise was considerable. Among the first scholars associated with Alexandria was Euclid, one of the most influential mathematicians of all times.

Euclid, about whose life nothing is known with any certainty, flourished probably during the time of the first Ptolemy (306–283 B.C.), to whom he is supposed to have remarked that "there is no royal road to geometry." His most famous and most advanced texts are the thirteen books of the *Elements* (*Stoicheia*), although he is also credited with several other minor texts. Among these other texts are the *Data*, containing what we would call applications of algebra to geometry but presented in strictly geometrical language. We do not know how many of these texts are Euclid's own and how many are compilations, but they show at many places an astonishing penetration. They are the first full mathematical texts that have been preserved.

The *Elements* form, next to the Bible, probably the most reproduced and studied book in the history of the Western World. More than a thousand editions have appeared since the invention of printing, and before that time manuscript copies dominated much of the teaching of geometry. Most of our school geometry was taken, often literally, from eight or nine of the thirteen books; and the Euclidean tradition still weighs heavily on our elementary instruction. For the professional mathematician these books have always had an inescapable fascination (even though their pupils often sighed) and their logical structure has influenced scientific thinking perhaps more than any other text in the world.

Euclid's treatment is based on a strictly logical deduction of theorems from a set of definitions, postulates, and axioms. The first four books deal with plane geometry but they do not deal with the theory of proportions and lead from the most elementary properties of lines and angles to the congruence of triangles, the equality of areas, the theorem of Pythagoras (Book I, Proposition 47), the construction of a square equal to a given rectangle, the golden section, the circle, and the regular polygons. Here the theorem of Pythagoras and the golden section are introduced as properties of areas. The fifth book presents Eudoxus' theory of proportions in its purely geometrical form, and in the sixth book this is applied to the similarity of plane figures. Here we return to the theorem of Pythagoras and the golden section (Book VI, Props. 31, 30), but now as theorems concerning ratios. This introduction of similarity at such a late stage is one of the most important differences between Euclid's presentation of plane geometry and the present one, and must be ascribed to the emphasis laid by Euclid on Eudoxus' novel theory of incommensurables. The geometrical discussion is resumed in the tenth book, often considered Euclid's most difficult one, which contains a geometrical classification of quadratic irrationals and their quadratic roots, hence of what we call numbers of the form $a \pm \sqrt{b}$, $\sqrt{a \pm \sqrt{b}}$. The last three books deal with solid geometry and lead via solid angles, the volumes of parallelepipeds, prisms, and pyramids to the sphere and to what seems to have been intended as the climax: the discussion of the five regular ("Platonic") bodies and the proof that only five such bodies exist.

Books VII–IX are devoted to number theory—not to computational technique but to such Pythagorean subjects as the divisibility of integers, the summation of the geometrical series, and some properties of prime numbers. There we find both "Euclid's algorithm" to find the greatest common divisor of a given set of numbers, and "Euclid's theorem" that there are an infinite number of primes (Book IX, Prop. 20). Of particular interest is the theorem (Book VI, Prop. 27) which contains the first maximum problem that has reached us, with the proof that the square, of all rectangles of given perimeter, has maximum area. The fifth postulate of Book I (the relation between "axioms" and "postulates" in Euclid is not clear) is equivalent to the so-called "parallel axiom," according to which one and only one line can be drawn through a point parallel to a given line. Attempts to reduce this axiom to a theorem led in the nineteenth century to a full appreciation of Euclid's wisdom in adopting it as an axiom and to the discovery of other, so-called non-Euclidean, geometries. Rejection of the axiom of Archimedes has in a similar way led to non-Archimedean geometries.

Algebraic reasoning in Euclid is cast entirely into geometric form. An expression $\sqrt{A}$ is introduced as the side of a square of area A, a product ab as the area of a rectangle with sides a and b. Linear and quadratic equations are solved by geometrical constructions leading to the so-called "application of areas." This mode of expression was primarily due to Eudoxus' theory of proportions, which consciously rejected numerical expressions for line segments and in this way dealt with incommensurables in a purely geometrical way—arithmetic confining itself only to "numbers" (integers) and their ratios.

What was Euclid's purpose in writing the *Elements?* We may assume with some confidence that he wanted to bring together into one text three great discoveries of the recent past: Eudoxus' theory of proportions, Theaetetus' theory of irrationals, and the theory of the five regular bodies which occupied an outstanding place in Plato's cosmology. These three were all typically Greek achievements.

8

The greatest mathematician of the Hellenistic period—and of antiquity as a whole—was Archimedes (287–212 B.C.), who lived in Syracuse as adviser to King Hiero. He is one of the few scientific figures of antiquity who is more than a name; several data about his life and person have been preserved. We know that he was killed when the Romans took Syracuse, after he placed his technical skill at the disposal of the defenders of the city. This interest in practical applications may appear odd if we compare it to the contempt in which such interest was held in the Platonic school of his contemporaries; however, an explanation is found in the much-quoted statement in Plutarch's *Vita Marcelli* (XVII, 4):

> . . . although these inventions had obtained for him the reputation of more than human sagacity, he did not deign to leave behind any written work on such subjects, but, regarding as ignoble and sordid the business of mechanics and every sort of art which is directed to use and profit, he placed his whole ambition in those speculations the beauty and subtlety of which are untainted by any admixture of the common needs of life.

This however, was written by a Platonist about three centuries after Archimedes. Authors who lived closer to his time, such as Polybius and Vitruvius, do not mention these pangs of conscience and see in Archimedes only the great master of mechanics.

The most important contributions which Archimedes made to mathematics were in the domain of what we now call the "integral calculus"—theorems on areas of plane figures and on volumes of solid bodies. In *Measurement of the Circle* he found an approximation of the circumference of the circle by the use of inscribed and circumscribed regular polygons. Extending his approximation to polygons of 96 sides he found (in our notation):

$$3\tfrac{10}{71} < 3\,\frac{284\tfrac{1}{4}}{2018\tfrac{7}{10}} < 3\,\frac{284\tfrac{1}{4}}{2017\tfrac{1}{4}} < \pi < 3\,\frac{667\tfrac{1}{2}}{4673\tfrac{1}{2}} < 3\,\frac{667\tfrac{1}{2}}{4672\tfrac{1}{2}} = 3\tfrac{1}{7}\;^{16}$$

which is usually expressed by saying that π is about equal to $3\tfrac{1}{7}$. In Archimedes' book *On the Sphere and Cylinder*, we find the expression for the area of the sphere (in the form that the area of a sphere is four times that of a great circle) and for the volume of a sphere (in the form that this volume is equal to 2/3 of the volume of the circumscribed cylinder). Archimedes' expression for the area of a parabolic segment (4/3 the area of the inscribed triangle with the same base as the segment and its vertex at the point where the tangent is parallel to the base) is found in his book *Quadrature of the Parabola*. In the book *On Spirals*, we find the "Spiral of Archimedes," with area computations; in *On Conoids and Spheroids* we find the volume of certain quadratic surfaces of revolution. Archimedes' name is also connected with his theorem on the loss of weight of bodies submerged in a liquid, which can be found in his book *On Floating Bodies*, a treatise on hydrostatics.

In all these works Archimedes combined a surprising originality of thought with a mastery of computational technique and rigor of demonstration. Typical of this rigor is the "Axiom of Archimedes" already quoted (see Sec. 5, above), and his consistent use of the exhaustion method to prove the results of his integration. We have seen how he actually found these results in a more heuristic way (by "weighing" infinitesimals); but he subsequently published them in accordance with the strictest requirements of rigor. In his computational proficiency Archimedes differed from most of the productive Greek mathematicians. This gave his work, with all its typically Greek characteristics, a touch of the Oriental. This touch is revealed in his "Cattle Problem," a very complicated problem in indeterminate analysis which may be interpreted as a problem leading to an equation of the "Pell" type:

$$t^2 - 4729494\,u^2 = 1,$$

which is solved by very large numbers.

This is only one of many indications that the Platonic tradition never entirely dominated Hellenistic mathematics; Hellenistic astronomy points in the same direction.

9

With the third great Hellenistic mathematician, Apollonius of Perga (c. 247–205 B.C.), we are again entirely within the Greek geometrical tradition. Apollonius, who seems to have taught at Alexandria and at Pergamum, wrote a treatise of eight books on *Conics*, seven of which have survived, three only in Arabic translation. It is a treatise on the ellipse, parabola, and hyperbola, introduced as sections

---

[16] $3.1409 < \pi < 3.1429$. The arithmetic mean of upper and lower limit gives $\pi = 3.1419$. The correct value is $3.14159\ldots$. The symbol π was not used in antiquity in our modern sense (in Ancient Greece, π meant 80). It was first used in that sense by William Jones (1706), a friend of Newton and the father of the great Sanskritist, but was generally adopted after Euler had used it in his *Introductio* of 1748.

σονα ἢ ὃν ‚γιγ L' δ' πρὸς ψπ. δίχα ἡ ὑπὸ ΓΑΗ τῇ
ΑΘ· ἡ ΑΘ ἄρα διὰ τὰ αὐτὰ πρὸς τὴν ΘΓ ἐλάσσονα
λόγον ἔχει ἢ ὃν ‚εⱵκδ L' δ' πρὸς ψπ ἢ ὃν ‚αωκγ
πρὸς σμ· ἑκατέρα γὰρ ἑκατέρας δ ιγ· ὥστε ἡ ΑΓ
5 πρὸς τὴν ΓΘ ἢ ὃν ‚αωλη θ ια' πρὸς σμ. ἔτι δίχα
ἡ ὑπὸ ΘΑΓ τῇ ΚΑ· καὶ ἡ ΑΚ πρὸς τὴν ΚΓ ἐλάσσονα [ἄρα] λόγον ἔχει ἢ ὃν ‚αξ πρὸς ξϛ· ἑκατέρα γὰρ
ἑκατέρας ιᾱ μ'· ἡ ΑΓ ἄρα πρὸς [τὴν] ΚΓ ἢ ὃν ‚αθ ϛ'
πρὸς ξϛ. ἔτι δίχα ἡ ὑπὸ ΚΑΓ τῇ ΛΑ· ἡ ΑΛ ἄρα
10 πρὸς [τὴν] ΛΓ ἐλάσσονα λόγον ἔχει ἢ ὃν τὰ ‚βιϛ ϛ'
πρὸς ξϛ, ἡ δὲ ΑΓ πρὸς ΓΔ ἐλάσσονα ἢ τὰ ‚βιζ δ'
πρὸς ξϛ. ἀνάπαλιν ἄρα ἡ περίμετρος τοῦ πολυγώνου
πρὸς τὴν διάμετρον μείζονα λόγον ἔχει ἤπερ ‚στλϛ
πρὸς ‚βιζ δ', ἅπερ τῶν ‚βιζ δ' μείζονά ἐστιν ἢ τρι-
15 πλασίονα καὶ δέκα οα'· καὶ ἡ περίμετρος ἄρα τοῦ
ϛϛγώνου τοῦ ἐν τῷ κύκλῳ τῆς διαμέτρου τριπλασίων
ἐστὶ καὶ μείζων ἢ ῑ οα'· ὥστε καὶ ὁ κύκλος ἔτι μᾶλ-
λον τριπλασίων ἐστὶ καὶ μείζων ἢ ῑ οα'.

ἡ ἄρα τοῦ κύκλου περίμετρος τῆς διαμέτρου τρι-
20 πλασίων ἐστὶ καὶ ἐλάσσονι μὲν ἢ ἑβδόμῳ μέρει, μείζονι δὲ ἢ ῑ οα' μείζων.

---

1 L'] Eutocius, γ' AB(C). 3 ‚εⱵκδ] Eutocius, e corr. B,
‚ετκδ ABC. L'] Eutocius, e corr. B, ε' A, 3̄ B. 4 σμ] B²C,
σν AB. ιγ'] B², ιγ α' A(C); δ ιγ om. B. 5 ια'] B², om.
AB(C). 7 ξϛ] C, e corr. B, σξϛ AB. 8 ἑκατέρας] B²,
εκατερα ABC. ιᾱ μ'· ἡ ΑΓ] B², Wallis, οιμαι AB, οιμ(·) C.
πρὸς ΓΚ Eutocius. ΚΓ ἢ ὃν] B², Wurmius; (ΓΚ) .. (χε)ν C,
καταγον A. ‚αθ ϛ'] B²C, ‚αοϛ A. 10 ΛΓ] Wallis, ΑΓ ABC;
πρὸς ΛΓ Eutocius. 13 ‚στλϛ] Eutocius, B², Wallis, ‚στᾱ ϛ'
ABC. 14 ‚βιζ (pr.)] e corr. B, ‚ξιζ AC. 15 οα'] B, corr.
ex ο' α' C, ο' ᾱ' A. 16 ϛϛγώνου] C, ϛϛ πολυγωνου AB. 17
ῑ οα'] e corr. B, ὃν ο' ια' AB(C). 18 ῑ οα'] e corr. B, θ' ια' AC.
20 ἐλάσσονι] scripsi, ελασσων ABC. μείζονι—21 μείζων] scripsi,

---

REPRODUCTION OF TWO PAGES OF "ARCHIMEDES OPERA OMNIA"
SHOWING COMPUTATION OF π.

(J. L. Heiberg, ed., Vol. I, 1910.) (This was not the way Archimedes
wrote it, since ancient Greece only used capital letters.)

$A\Gamma : \Gamma H < 3013\frac{1}{2} \frac{1}{4} : 780$ [u. Eutocius].

secetur $\angle \Gamma A H$ in duas partes aequales recta $A\Theta$; propter eadem igitur erit $A\Theta : \Theta\Gamma < 5924\frac{1}{2}\frac{1}{4} : 780$ [u. Eutocius] siue $< 1823 : 240$; altera[1]) enim alterius $\frac{4}{13}$ est [u. Eutocius]; quare $A\Gamma : \Gamma\Theta < 1838\frac{9}{11} : 240$ [u. Eutocius]. porro secetur $\angle \Theta A\Gamma$ in duas partes aequales recta KA; est igitur $AK : K\Gamma < 1007 : 66$ [u. Eutocius]; altera[1]) enim alterius est $\frac{11}{40}$; itaque

$A\Gamma : \Gamma K < 1009\frac{1}{6} : 66$ [u. Eutocius].

porro secetur $\angle KA\Gamma$ in duas partes aequales recta $\varLambda A$; erit igitur

$\varLambda A : \varLambda\Gamma < 2016\frac{1}{6} : 66$ [u. Eutocius],

et $A\Gamma : \Gamma\varLambda < 2017\frac{1}{4} : 66$ [u. Eutocius]. et e contrario $\langle \Gamma\varLambda : A\Gamma > 66 : 2017\frac{1}{4}$ [Pappus VII, 49 p. 688]. sed $\Gamma\varLambda$ latus est polygoni 96 latera habentis; quare$\rangle$[2]) perimetrus polygoni ad diametrum maiorem rationem habet quam $6336 : 2017\frac{1}{4}$, quae maiora sunt quam triplo et $\frac{10}{71}$ maiora quam $2017\frac{1}{4}$; itaque etiam perimetrus polygoni inscripti 96 latera habentis maior est quam triplo et $\frac{10}{71}$ maior diametro; quare etiam multo magis[3]) circulus maior est quam triplo et $\frac{10}{71}$ maior diametro.

ergo ambitus circuli triplo maior est diametro et excedit spatio minore quam $\frac{1}{7}$, maiore autem quam $\frac{10}{71}$.

---

1) Exspectaueris ἑκάτερος (sc. ὅρος) γὰρ ἑκατέρου (ἑκάτερα γὰρ ἑκατέρων Wallis), sed genus femininum minus adcurate refertur ad auditum uerbum εὐθεῖα, quasi sit $A\Theta = 5924\frac{1}{2}\frac{1}{4}$, $\Theta\Gamma = 780$.

2) Ueri simile est, Archimedem ipsum haec addidisse. similes omissiones durae inueniuntur p. 240, 4, 6; 242, 5, 8, nec dubito eas transscriptori tribuere, sicut etiam p. 240, 8 τὸ πολύγωνον pro ἡ περίμετρος τοῦ πολυγώνου, p. 242, 17 ὁ κύκλος pro ἡ περίμετρος τοῦ κύκλου.

3) Quippe quae maior sit perimetro polygoni (De sph. et cyl. I p. 10, 1).

μείζων δε AC, maior B, autem quam decem septuagesimunis add. B². In fine: Ἀρχιμήδους κύκλου μετρήσις A.

of a circular cone, and penetrates as far as the discussion of the evolutes of a conic. We know these conics by the names found in Apollonius; they refer to certain area properties of these curves, which are expressed in our notation by the equations (homogeneous notation, p, d are lines in Apollonius)

$$y^2 = px; \; y^2 = px \pm \frac{p}{d}x^2.$$

(The plus gives the hyperbola, the minus the ellipse.) Parabola here means "application"; ellipse, "application with deficiency"; hyperbola, "application with excess." Apollonius did not have our coordinate method because he had no algebraic notation (probably rejecting it consciously under influence of the school of Eudoxus). Many of his results, however, can be transcribed immediately into coordinate language—including his property of the evolutes, which is identical with the Cartesian equation.[17] This can also be said for other books by Apollonius, parts of which have been preserved and which contain "algebraic" geometry in geometrical and, therefore, homogeneous language. Here we find Apollonius' *tangency problem*, which requires the construction of the circles tangent to three given circles; the circles may be replaced by straight lines or points. In Apollonius we meet for the first time in explicit form the requirement that geometrical constructions be confined to the compass and ruler only; therefore it was not such a general Greek requirement as is sometimes believed.

10

Mathematics, throughout history until modern times, cannot be separated from astronomy. The needs of irrigation and of agriculture in general—and to a certain extent also of navigation—accorded to astronomy the first place in Oriental and Hellenistic science, and its course determined to no small extent that of mathematics. The computational and often the conceptual content of mathematics was largely conditioned by astronomy, and the progress of astronomy depended equally on the power of the mathematical books available. The structure of the planetary system is such that relatively simple mathematical methods allow far-reaching results, but are at the same time complicated enough to stimulate improvement of these methods and of the astronomical theories themselves. The Orient itself had made considerable advances in computational astronomy during the period just preceding the Hellenistic era, especially in Mesopotamia during the late Assyrian and Persian periods. Here observations consistently carried on over a long period had allowed a remarkable understanding of many ephemerides. The motion of the moon was one of the most challenging of all astronomical problems to the mathematician, in antiquity as well as in the eighteenth century, and Babylonian (Chaldean) astronomers devoted much effort to its study. The meeting of Greek and Babylonian science during the Seleucid period brought

---

[17]"My thesis, then, is that the essence of analytic geometry is the study of *loci* by means of their equations, and that this was known to the Greeks and was the basis of their study in conic sections." J. L. Coolidge, *A History of Geometrical Methods* (Oxford, 1940), p. 119. See, however, our remarks on Descartes. Coolidge's remarks strike us as unhistoric: the whole way of mathematical thinking of these Greeks was different from ours.

great computational and theoretical advancement, and where Babylonian science continued in the ancient calendric tradition, Greek science produced some of its most significant theoretical triumphs.

The oldest known Greek contribution to theoretical astronomy was the planetary theory of the same Eudoxus who inspired Euclid. It was an attempt to explain the motion of the planets (around the earth) by assuming the superposition of four rotating concentric spheres, each with its own axis of rotation with the ends fixed on the enclosing sphere. This was something new, and typically Greek, an explanation rather than a chronicle of celestial phenomena. Despite its special form, Eudoxus' theory contained the central idea of all planetary theories until the seventeenth century; it consisted in the explanation of irregularities in the apparent orbits of moon and planets by the superposition of circular movements. It still underlies the computational side of our modern dynamic theories as soon as we introduce Fourier series.

Eudoxus was followed by Aristarchus of Samos (c. 280 B.C.), the "Copernicus of antiquity," credited by Archimedes with the hypothesis that the sun, and not the earth, is the center of the planetary motion. This hypothesis found few adherents in antiquity, although the belief that the earth rotates about its axis had a wide acceptance. The small success of the heliocentric hypothesis was mainly due to the authority of Hipparchus, often considered the greatest astronomer of antiquity.

Hipparchus of Nicaea made observations between 141 and 127 B.C. Little of his work has come to us directly, the main source of our knowledge of his achievements coming through Ptolemy, who lived three centuries later. Much of the contents of Ptolemy's great work, the *Almagest*, may be ascribed to Hipparchus, especially the use of eccentric circles and epicycles to explain the motion of sun, moon, and planets, as well as the discovery of the precession of the equinoxes. Hipparchus is also credited with a method to determine latitude and longitude by astronomical means, but antiquity never was able to muster a scientific organization sufficient to do any large-scale mapping. (Scientists in antiquity were very thinly scattered, both in locality and in time.) Hipparchus' work was closely connected with the achievements of Babylonian astronomy, which reached great heights in this period; and we may see in this work the most important scientific fruit of the Greek-Oriental contact of the Hellenistic period.[18]

11

The third and last period of antique society is that of the Roman domination. Syracuse fell to Rome in 212, Carthage in 146, Greece in 146, Mesopotamia in 64, and Egypt in 30 B.C. The entire Roman-dominated Orient, including Greece, was reduced to the status of a colony ruled by Roman administrators. This control did not affect the economic structure of the Oriental countries as long as the heavy taxes and other levies were duly delivered. The Roman Empire naturally split into a Western part with extensive agriculture fitted for wholesale slavery, and an

[18]O. Neugebauer, *Exact Science in Antiquity*, Univ. of Pennsylvania Bicentennial Conf., Studies in Civilization (Philadelphia, 1941), pp. 22–31; and the same author's *The Exact Sciences in Antiquity* (Princeton, 1952; 2nd ed., 1957; Dover reprint, 1969).

Eastern part with intensive agriculture which never used slaves except for domestic duties and for public works. Despite the growth of some cities and a commerce embracing the whole of the known Western World, the entire economic structure of the Roman Empire remained based on agriculture. The spread of a slave economy in such a society was fatal to all original scientific work. Slave owners as a class were seldom interested in technical discoveries, partly because slaves could do all the work cheaply, and partly because they feared to place any tool in the hands of slaves which might sharpen their intelligence. There were, however, also learned slaves. Whoever wanted serious learning in philosophy and science studied Greek. Many members of the ruling class thus dabbled in the arts and sciences, but this very dabbling promoted mediocrity rather than productive thinking. When with the decline of the slave market the Roman economy declined, there were few men to cultivate even the mediocre science of the past centuries.

As long as the Roman Empire showed some stability, Eastern science continued to flourish as a curious blend of Hellenistic and Oriental elements. Although originality and stimulation gradually disappeared, the *pax Romana*, lasting for many centuries, allowed undisturbed speculation along traditional lines. Coexistent with the *pax Romana* for some centuries was the *pax Sinensis*; in all its history the Eurasian continent never knew such a period of uninterrupted peace as under the Antonines in Rome and the Han in China. This made the diffusion of knowledge over the continent from Rome and Athens to Mesopotamia, China, and India easier than ever before. Hellenistic science continued to flow into Iran and India and was influenced in turn by the science of these countries. Glimpses of Babylonian astronomy and Greek mathematics came to Italy, Spain, and Gaul—an example is the spread of the sexagesimal division of angle and hour over the Roman Empire. There exists a theory of F. Woepcke which traces the spread of the so-called Hindu-Arabic numerals throughout Europe to neo-Pythagorean influences in the later Roman Empire (1863). This spread at that time may or may not be true, but if the event dates back that far, then it is more likely due to the influences of trade rather than of philosophy.

Alexandria remained the center of antique mathematics. Original work continued, although compilation and commentarization became more and more the prominent form of science. Many results of the ancient mathematicians and astronomers have been transmitted to us through the works of these compilers; and it is sometimes quite difficult to find out what they transcribed and what they discovered themselves. In trying to understand the gradual decline of Greek mathematics we must also take into account its technical side: the clumsy geometrical mode of expression with the consistent rejection of algebraic notation, which made any advance beyond the conic sections almost impossible. Algebra and computation were left to the despised Orientals, whose lore was covered by a veneer of Greek civilization. It is wrong, however, to believe that Alexandrian mathematics was purely "Greek" in the traditional Euclidean-Platonic sense; computational arithmetic and algebra of an Egyptian-Babylonian type were cultivated side by side with abstract geometrical demonstrations. We have only to think of Ptolemy, Heron (Hero), and Diophantus to become convinced of this fact. The only tie between the many races and schools was the common use of Greek.

12

One of the earliest Alexandrian mathematicians of the Roman period was Nicomachus of Gerasa (c. A.D. 100), whose *Introduction to Arithmetic* is the most complete exposition extant of Pythagorean arithmetic. It deals in great part with the same subjects as the arithmetical books of Euclid's *Elements*, but where Euclid represents numbers by straight lines, Nicomachus uses arithmetical notation with ordinary language when undetermined numbers are expressed. His treatment of polygonal numbers and pyramidal numbers influenced medieval arithmetic, especially through Boethius.

One of the greatest documents of this second Alexandrian period was Ptolemy's *Great Collection*, better known under the Arabicized title of *Almagest* (c. A.D. 150). The *Almagest* was an astronomical opus of supreme mastership and originality, even though many of the ideas may have come from Hipparchus or Kidinnu and other Babylonian astronomers. Also, it contained a trigonometry, with a table of chords belonging to different angles ascending by halves of an angle, equivalent to a sine table according to the formula: chord $\alpha = 2R \sin \alpha/2$, where $R = 60$. Ptolemy found for the chord of 1° the value (1, 2, 50) $= \frac{1}{60} + \frac{2}{60^2} + \frac{50}{60^3} = .017453$; for π the value (3, 8, 30) $= \frac{377}{120} = 3.14166$. We find in the *Almagest* the formula for the sine and cosine of the sum and difference of two angles, together with a beginning of spherical trigonometry. The theorems were expressed in geometrical form—our present trigonometrical notation dates only from Euler in the eighteenth century. We also find in this book "Ptolemy's theorem" for a quadrilateral inscribed in a circle. In Ptolemy's *Planisphaerium* we find a discussion of stereographic projection; in his *Geographia* the position of places on earth is determined by latitude and longitude, which are ancient examples of coordinates on the sphere. Stereographic projection underlies the construction of the astrolabe, an instrument used for the determination of position on earth, already known in antiquity and widely used until the introduction of the octant, later sextant, in the eighteenth century.[19]

Somewhat older than Ptolemy was Menelaus (c. A.D. 100), whose *Sphaerica* contained a geometry of the sphere, with a discussion of spherical triangles, a subject which is missing in Euclid. Here we find "Menelaus' theorem" for the triangle in its extension to the sphere. Whereas Ptolemy's astronomy contained a good deal of computational work in sexagesimal fractions, Menelaus' treatise was geometrical in the pure Euclidean tradition.

To the period of Menelaus may also belong Heron (Hero); at any rate, we know

---

[19]H. Michel, *Traité de l'astrolabe* (Paris, 1947). Also, O. Neugebauer, "The Early History of the Astrolabe," *Isis*, Vol. 40 (1949), pp. 240–56. More general: Eva G. L. Taylor, *The Haven-finding Art* (1956).

that he described accurately a lunar eclipse of A.D. 62.[20] Under his name a number of books have come down to us on geometrical, computational, and mechanical subjects; this writing shows a curious blend of the Greek and the Oriental. In *Metrica* he derived the "Heronic formula" for the area of a triangle,

$$A = \sqrt{s(s - a)(s - b)(s - c)},$$

in purely geometrical form; the proposition itself has been ascribed to Archimedes. Also in *Metrica* we find typical Egyptian unit fractions, as in the approximation of $\sqrt{63}$ by

$$7 + \tfrac{1}{2} + \tfrac{1}{4} + \tfrac{1}{8} + \tfrac{1}{16}.$$

Heron's formula for the volume of a frustum of a square pyramid can readily be reduced to the one found in the ancient Moscow papyrus. His measurement of the volume of the five regular polyhedra, on the contrary, was again in the spirit of Euclid.

13

The Oriental touch is even stronger in the *Arithmetica* of Diophantus (c. A.D. 250). Only six of the original books survive; the total number is a matter of conjecture. Their skillful treatment of indeterminate equations shows that the ancient algebra of Babylon or perhaps India not only survived under the veneer of Greek civilization but also was improved by a few active men. How and when it was done is not known, just as we do not know who Diophantus was—he may have been a Hellenized Babylonian. His book is one of the most fascinating treatises preserved from Greco-Roman antiquity.

Diophantus' collection of problems is of wide variation, and their solution is often highly ingenious. "Diophantine analysis" consists in finding answers to indeterminate equations such as

$$Ax^2 + Bx + C = y^2$$

$$Ax^3 + Bx^2 + Cx + D = y^2,$$

or sets of these equations. Typical of Diophantus is that he was only interested in positive rational solutions; he called irrational solutions "impossible," and was careful to select his coefficients so as to get the positive rational solution he was looking for. Among the equations we find

$$x^2 - 26y^2 = 1$$

$$x^2 - 30y^2 = 1,$$

---

[20]O. Neugebauer, "Über eine Methode zur Distanzbestimmung Alexandria–Rom bei Heron," *Hist. fil. Medd. Danske Vid. Sels.*, Vol. 26, No. 2 (1938), 28pp.

now known as the "Pell" equations. Diophantus also has several propositions in the theory of numbers, such as the theorem (III, 19) that if each of two integers is the sum of two squares their product can be resolved in two ways into two squares. There are also theorems about the division of a number into the sum of three and four squares.

In Diophantus we find the first systematic use of algebraic symbols. He has a special sign for the unknown, for the minus, for reciprocals. The signs are still of the nature of abbreviations rather than algebraic symbols in our sense (they form the so-called "syncopated" algebra); for each power of the unknown there exists a special symbol.[21] There is no doubt that we have here not only, as in Babylon, arithmetical questions of a definite algebraic nature, but also a well-developed algebraic notation which was greatly conducive to the solution of problems of greater complexity than were ever taken up before.

14

The last of the great Alexandrian mathematical treatises was written by Pappus (early fourth century). His *Collection* (*Synagoge*) was a kind of handbook to the study of Greek geometry with historical annotation, improvements, and alterations of existing theorems and demonstrations. It was to be read with the original works rather than independently. Many results of ancient authors are known only in the form in which Pappus preserved them. Examples are the problems dealing with the quadrature of the circle, the duplication of the cube, and the trisection of the angle. Interesting is his chapter on isoperimetric figures, in which we find the circle has a larger area than any regular polygon of equal perimeter. Here is also the remark that the cells in the honeycomb satisfy certain maximum-minimum properties.[22] Archimedes' semiregular solids are also known through Pappus. Like Diophantus' *Arithmetica*, the *Collection* is a challenging book whose problems inspired much further research in later days.

The Alexandrian school gradually died with the decline of antique society. It remained, as a whole, a bulwark of paganism against the progress of Christianity, and several of its mathematicians also left traces in the history of ancient philosophy. Proclus (410–485), whose *Commentary on the First Book of Euclid* is one of our main sources of the history of Greek mathematics, headed a Neoplatonic school in Athens. Another representative of this school, in Alexandria, was Hypatia, who wrote commentaries on the classical mathematicians. She was murdered in 415 by the followers of St. Cyril, a fate which inspired a novel by Charles Kingsley.[23] These philosophical schools with their commentators had

---

[21]Papyrus 620 of the University of Michigan, acquired in 1921, contains some problems in Greek algebra dating to a period before Diophantus, perhaps early second century A.D. Some symbols found in Diophantus appear in this manuscript. See F. E. Robbins, *Classical Philology*, Vol. 24 (1929), pp. 321–29; K. Vogel, *ibid.*, Vol. 25 (1930), pp. 373–75. The distinction between rhetoric algebra (all words), syncopated algebra, and symbolic algebra (our present one) is due to G.H.F. Nesselman, *Die Algebra der Griechen* (Berlin, 1842).

[22]There is a full discussion of this problem in D'Arcy W. Thompson, *Growth and Form*, 2nd ed. (Cambridge, 1942). With the study of these so-called isoperimetric problems, the name of Zenodorus (second century B.C.) is connected. This treatise has been lost, but it is summarized by Pappus. Compare Steiner (see Chapter VIII, Section 9, below).

[23]C. Kingsley, *Hypatia* (1853). Also F. Mauthner, *Hypatia* (1892, in German).

their ups and downs for centuries. The Academy in Athens was discontinued as "pagan" by the Emperor Justinian (529), but by this time there were again schools in such places as Constantinople and Jundīshāpūr. Many old codices survived in Constantinople, while commentators continued to perpetuate the memory of Greek science and philosophy in the Greek language. In 630 Alexandria was taken by the Arabs, who replaced the upper layer of Greek civilization in Egypt by an upper layer of Arabic. There is no reason to believe that the Arabs destroyed the famous Alexandrian Library, since it is doubtful whether this library still existed at that time.[24] As a matter of fact, the Arabic conquests did not materially change the character of mathematical studies in Egypt. There may have been a retrogression, but when we hear of Egyptian mathematics again it is still following the ancient Greco-Oriental tradition (e.g. Alhazen).

15

We end this chapter with some remarks on Greek arithmetic and logistics. Greek mathematicians made a distinction between "arithmetica" or science of numbers (*arithmoi*) and "logistics" or practical computation. The term *arithmos* expressed only a natural number, a "quantity composed of units" (Euclid, Book VII, Def. 2; this also meant that "one" was not considered a number).[25] Our conception of real number was unknown. A line segment, therefore, did not always have a length. Geometrical reasoning replaced our work with real numbers. When Euclid wanted to express that the area of a triangle is equal to half the base times the altitude, he had to state that it is half the area of a parallelogram of the same base and lying between the same parallels (Euclid I, 41). Pythagoras' theorem was a relation between the areas of three squares and not between the lengths of three sides. Euclid's *Elements* has a theory of quadratic equations, but it is expressed in terms of a so-called application of areas, and since the roots are line segments obtained by performing certain indicated constructions, we can say that the only roots admitted are positive ones. However, in the *Elements* a line segment does not necessarily have a numerical value attached to it. These concepts of lines and numbers must be considered as a deliberate act brought about by the victory of Platonic idealism among those sections of the Greek ruling class interested in mathematics, since the contemporary Oriental conceptions concerning the relation of algebra and geometry did not admit any restriction of the number concept. There is every reason to believe that, for the Babylonians, Pythagoras' theorem was a numerical relation between the lengths of sides, and it was this type of mathematics with which the Ionian mathematicians had become acquainted.

Ordinary computational mathematics, known as "logistics," remained very much alive during all periods of Greek history. Euclid rejected it, but Archimedes and Heron used it with ease and without scruples. Actually it was based on a system of numeration which changed with the times. The early Greek method of numeration was based on an additive decimal principle like that of the Egyptians

---

[24]E. A. Parsons (*The Alexandrian Library. Glory of the Hellenistic World* [Amsterdam, etc., 1952]) claims the opposite.

[25]This lasted until the Renaissance. Stevin, in his *Arithmétique* of 1585, pleads passionately for the recognition of "one" as a number like other integers.

and the Romans. In Alexandrian times, or perhaps earlier, a method of writing numbers appeared which was used for fifteen centuries, not only by scientists but also by merchants and administrators. It used the successive symbols of the Greek alphabet to express, first our symbols 1, 2, . . ., 9, then the tens from 10 to 90, and finally the hundreds from 100 to 900 ($\alpha = 1$, $\beta = 2$, etc.). Three extra archaic letters were added to the 24 letters of the Greek alphabet in order to obtain the necessary 27 symbols. With the aid of this system every number less than 1000 could be written with at most three symbols, e.g., 14 as $\iota\delta$, since $\iota = 10$, $\delta = 4$; numbers larger than 1000 could be expressed by a simple extension of the system. This system is used in the existing manuscripts of Archimedes, Heron, and all the other classical authors. There is archaeological proof that it was taught in the schools.

This was a decimal nonposition system, both $\iota\delta$ and $\delta\iota$ could only mean 14. This lack of place value and the use of no less than 27 symbols have occasionally been taken as a proof of the inferiority of the system. The ease with which the ancient mathematicians used it, its acceptance by Greek merchants even in rather complicated transactions, and its long persistence—in the East Roman Empire until its very end in 1453—seem to point to certain advantages. Some practice with the system can indeed convince us that it is possible to perform the four elementary operations easily enough once the meaning of the symbols is mastered. Fractional calculus with a proper notation is also simple; but the Greeks were inconsistent because of their lack of a uniform system. They used Egyptian unit fractions, Babylonian sexagesimal fractions, and also fractions in a notation reminiscent of ours. Decimal fractions were never introduced, but this great improvement appears only late in the European Renaissance after the computational apparatus had extended far beyond anything ever used in antiquity; even then decimal fractions were not adopted in many schoolbooks until the eighteenth and nineteenth centuries.

It has been argued that this alphabetical system was detrimental to the growth of Greek algebra, since the use of letters for definite numbers prevented their use for denoting numbers in general, as we do in our algebra. Such a formal explanation of the absence of a Greek algebra before Diophantus must be rejected, even if we accept the great value of an appropriate notation. If the classical authors had been interested in algebra, they would have created the appropriate symbolism, with which Diophantus actually made a beginning. The problem of Greek algebra can be elucidated only by further study of the connections between Greek mathematicians and Babylonian algebra in the framework of the entire relationship of Greece and the Orient.

Literature

The Greek classical authors are available in excellent texts, with the principal texts also available in English translations. The best introduction is offered in the following books:

Heath, T. L. *A History of Greek Mathematics*. 2 vols., Cambridge, 1921. Dover reprint, 1981.
——. *A Manual of Greek Mathematics*. Oxford, 1931.
——. *The Thirteen Books of Euclid's Elements*. 3 vols., Cambridge, 1908. Dover reprint, 1956.
Ver Eecke, P. *Oeuvres complètes d'Archimède*. Brussels, 1921.
——. *Pappus d'Alexandrie. La Collection mathématique*. Paris-Bruges, 1933.
——. *Proclus de Lycie. Les commentaires sur le premier livre des éléments d'Euclide*. Bruges, 1948.
Scriba, C. *Archimedes*.
Dijksterhuis, E. J. *Archimedes*. Copenhagen, 1956.
Manitius, K. *Ptolemäus' Handbuch der Astronomie*, 2nd ed., O. Neugebauer, ed. 2 vols., Leipzig, 1963 (1st ed., Leipzig, 1912–13).
Pedersen, O. *A Survey of the Almagest*. Copenhagen, 1974.
[Toomer, G. J., ed.] *Diocles on Burning Mirrors*. Berlin, etc., 1976.
Bashmakova, I. G. *Diophantus und diophantische Gleichungen*, 1972 (from the Russian). See also *Istor.-mat. Issled*. Vol. 17 (1966), pp. 185–204.
Loria, G. *Le Scienze esatte nell'antica Grecia*, 2nd ed. Milan, 1914.
Allman, G. J. *Greek Geometry from Thales to Euclid*. Dublin, 1889.
Gow, J. *A Short History of Greek Mathematics*. Cambridge, 1884.
Reidemeister, K. *Die Arithmetik der Griechen*. Hamburg Math. Seminar Einzelschriften 26, 1939.
——. *Das exakte Denken der Griechen*. Hamburg, 1949.
Cohen, M. R., and Drabkin, I. E. *A Source Book in Greek Science*. London, 1948.
Heath, T. L. *Mathematics in Aristotle*. Oxford, 1949.
van der Waerden, B. L. *Ontwakende Wetenschap*. Groningen, 1950. English trans., *Science Awakening*, Oxford, 1961. (Deals with Egyptian, Babylonian, and Greek mathematics.)
Becker, O. *Das mathematische Denken der Antike*. Göttingen, 1957.
Hauser, G. *Geometrie der Griechen von Thales bis Euclid*. Lucerne, 1955.
Blaschke, W. *Griechische und anschauliche Geometrie*. Munich, 1953.
Dantzig, T. *The Bequest of the Greeks*. New York, 1955.
Kolman, E. *History of Mathematics in Antiquity*. Moscow, 1961 (in Russian).
Wussing, H. *Mathematik in der Antike*. Leipzig, 1965. (The books of Kolman and Wussing also deal with Egyptian and Babylonian mathematics.)
Steele, A. D. "Über die Rolle von Zirkel und Lineal in der griechischen Mathematik." *Quellen und Studien*, Vol. A2 (1932), pp. 61–89.

Comparative Greek, Latin, and English texts are in:

Thomas, I. *Selections Illustrating the History of Greek Mathematics*. Cambridge, Mass., and London, 1939.

Further text criticism is given in:

Tannery, P. *Pour l'histoire de la science hellène*, 2nd ed. Paris, 1930.
——. *Mémoires scientifiques*. Vols. 1–4, Toulouse and Paris, 1912–20.
Vogt, H. "Die Entdeckungsgeschichte des Irrationalen nach Plato und anderen Quellen des 4ten Jahrhunderts." *Bibliotheca mathematica*, Vol. 3, No. 10 (1909–10), pp. 97–105.

Sachs, E. *Die fünf Platonischen Körper.* Berlin, 1917.

Hellen, S. "Die Entdeckung der stetigen Teilung durch die Pythagoreen," *Abh. Deut. Akad. Wiss., Kl. f. Math. u. Physik u. Techn.*, No. 6 (1958).

Frank, E. *Plato und die sogenannten Pythagoreer.* Halle, 1923.

Apostle, H. O. *Aristotle's Philosophy of Mathematics.* Chicago, 1952.

Luria, S. "Die Infinitesimaltheorie der antiken Atomisten." *Quellen und Studien*, Vol. B 2 (1932), pp. 106–85.

Lorenzen, P. *Die Entstehung der exakten Wissenschaften.* Berlin, 1960.

Szabo, A. *The Beginnings of Greek Mathematics.* Dordrecht, 1978. (Trans. from the German, Munich and Vienna, 1969.) See also *HM*, Vol. 1 (1974), pp. 291–316.

Knorr, W. R. "Archimedes and the Pre-Euclidean Proportion Theory." *Arch. intern. hist. sc.*, Vol. 28 (1978), pp. 183–244.

A good critical survey of the various hypotheses concerning Greek mathematics is given in:

Dijksterhuis, E. *De elementen van Euclides.* 2 vols., Groningen, 1930 (in Dutch).

See also the article in *DSB* on Euclid, as well as:

van der Waerden, B. L. *Die Pythagoreër. Religiöse Brüderschaft und Schule der Wissenschaft.* Zurich and Munich, 1979.

Lexikon der Antiken Welt. Stuttgart, 1965. Contains articles on mathematics by K. von Fritz, H. Guericke, K. Vogel.

On Byzantine mathematics, see:

Vogel, K. "Der Anteil von Byzanz an Erhaltung und Weiterverbreitung der griechischen Mathematik." In *Miscellanea mediaevalea Ia.* Berlin, 1962, pp. 112–28. (See *Istor.-mat. Issled.*, Vol. 10 [1973], pp. 249–63.)

CHAPTER IV

The Orient After the Decline of Greek Society

1

The ancient civilization of the Near East never disappeared despite all Hellenistic influence. Both Oriental and Greek influences are clearly revealed in the science of Alexandria; Constantinople and India were also important meeting grounds of East and West. In 394 Theodosius I founded the Byzantine Empire; its capital, Constantinople, was Greek, but it was the center of administration of vast territories where the Greeks were only a fraction of the urban population. For a thousand years this empire fought against forces from the East, North, and West, serving at the same time as a guardian of Greek culture and as a bridge between the Orient and the Occident. Mesopotamia became independent of the Romans and Greeks as early as the second century A.D., first under the Parthian kings, later (266) under the purely Persian dynasty of the Sassanians. The Indus region had for some centuries several Greek dynasties, which disappeared by the first century A.D.; but the native Indian kingdoms which followed kept up cultural relations with Persia and the West.

The political hegemony of the Greeks over the Near East disappeared almost entirely with the sudden growth of Islam. After 622, the year of the Hegira, the Arabs conquered large sections of Western Asia in an amazing sweep and in less than a century had occupied parts of the West Roman empire as far as Sicily, North Africa, and Spain. Wherever they went they tried to replace Greco-Roman civilization with Islamic. The official language became Arabic, instead of Greek or Latin; but the fact that a new language was used for scientific documents tends to obscure the truth that under Arabic rule a considerable continuity of culture remained. The ancient native civilizations had even a better chance to survive under this rule than under the alien rule of the Greeks. Persia, for instance, remained very much the ancient country of the Sassanians, despite the Arabic administration. However, the contest between the different traditions continued, only now in a new form. Throughout the whole period of Islamic rule there existed a definite Greek tradition holding its own against the different native cultures.

2

We have seen that the most glorious mathematical results of this competition and blending of Oriental and Greek culture during the heyday of the Roman Empire

65

peared in Egypt. With the decline of the Roman Empire the center of mathe-matical research began to shift to India and later back to Mesopotamia. The first well-preserved Indian contributions to the exact sciences are the *Siddhāntās*, of which one, the *Sūrya*, may be extant in a form resembling the original one (*c.* A.D. 300–400). These books deal mainly with astronomy and operate with epicycles and sexagesimal fractions. These facts suggest the influence of Greek astronomy, perhaps transmitted in a period antedating the *Almagest*; they also may indicate direct contact with Babylonian astronomy. In addition the *Siddhāntās* show many native Indian characteristics. The *Sūrya Siddhāntā* has tables of sines (*jyā*) in-stead of chords. This sine is a half-chord, hence a line segment.

The results of the *Siddhāntās* were systematically explained and extended by schools of Indian mathematicians, mainly centered in Ujjain (Central India) and Mysore (S. India). From the fifth century A.D., names and books of individual Indian mathematicians have been preserved; some books are even available in English translations.

The best known of these mathematicians are Āryabhata (called "the first," *c.* 500) and Brahmagupta (*c.* 625). The whole question of their indebtedness to Greece, Babylon, and China is a subject of much conjecture; but at the same time they show considerable originality. Characteristic of their work are the arith-metical-algebraic parts, which in their love for indeterminate equations bear some kinship to Diophantus. These writers were followed in the next centuries by others working in the same general field; their works were partly astronomical, partly arithmetical-algebraic, and had excursions into mensuration and trigonome-try. Āryabhata I had for π the value 3.1416. A favorite subject was the finding of rational triangles and quadrilaterals, in which Mahāvīra of the Mysore school (*c.* 850) was particularly prolific. Around 1150 we find in Ujjain, where Brahmagupta had worked, another excellent mathematician, Bhāskara. The first general solu-tion of indeterminate equations of the first degree $ax + by = c$ (where a, b, c are integers) is found in Brahmagupta. It is therefore, strictly speaking, incorrect to call linear indeterminate equations Diophantine equations. Where Diophantus still accepted fractional solutions, the Hindus were only satisfied with integer solu-tions. They also advanced beyond Diophantus in admitting negative roots of equations, though this may again have been an older practice suggested by Babylonian astronomy. Bhāskara, for instance, solved $x^2 - 45x = 250$ by $x = 50$ and $x = -5$; he indulged in some skepticism as to the validity of the negative root. His *Lilāvati* was for many centuries a standard work on arithmetic and mensuration in the East; the emperor Akbar had it translated into Persian (1587) and an English translation of 1817 was republished in 1827 in Calcutta. The original Sanskrit edition has often been republished: there exists one with Sanskrit and Hindi commentaries (Benares, 1961).[1] Ancient India still yields many more mathe-

---

[1]Brahmagupta states somewhere in his book that some of his problems were proposed "simply for pleasure." This confirms the fact that mathematics in the Orient had long since evolved from its purely utilitarian function, if that had ever been its sole one. One hundred and fifty years later Alcuin, in the West, wrote *Problems for the Quickening of the Mind of the Young*, expressing a similar nonutilitarian purpose. Mathematics in the form of the intellectual puzzle has often contrib-uted essentially to the progress of science by opening new fields. Some puzzles still await their integration into the main body of mathematics.

matical treasures; we now know, for instance, that the Gregory-Leibniz series for $\pi/4$ can already be found in a Sanskrit manuscript ascribed to Nīlakantha (c. 1500).[2]

3

The best-known achievement of Hindu mathematics is our present decimal position system. The decimal system is very ancient, and so is the position system; but their combination appears in China and then in India, where in the course of time it was gradually imposed upon older nonposition systems. Its first Indian occurrence is on a plate of the year 595, where the date 346 is written in decimal place-value notation. The Hindus long before this epigraphic record had a system of expressing large numbers by means of words arranged according to a place-value method. There are early texts in which the word śūnya, meaning zero, is explicitly used.[3] The so-called Bakshāli manuscript, consisting of seventy leaves of birch bark of uncertain origin and date (estimates range from the third to the twelfth century A.D.), and with traditional Hindu material on indeterminate and quadratic equations as well as approximations, has a dot to express zero. The oldest epigraphic record with a sign for zero dates from the ninth century. This is all much later than the occurrence of a sign for zero in Babylonian texts. The sign 0 for zero may be due to Greek influence (ouden = Greek for "nothing"); whereas the Babylonian dot only appears between digits, the Indian zero also appears at the end, so that 0, 1, 2, . . ., 9 become equivalent digits.[4]

The decimal place-value system slowly penetrated, probably originally from China, along the caravan roads into many parts of the Near East, taking its place beside other systems. Penetration into Persia, perhaps also Egypt, may very well have happened in the Sassanian period (224–641), when the contact between Persia, Egypt, and India was close. In this period the memory of the ancient Babylonian place-value system may still have been alive in Mesopotamia. The oldest definite reference to the Hindu place-value system outside of India is found in a work of 662 written by Severus Sēbōkht, a Syrian bishop. With Al-Fazārī's translation of the Siddhāntās into Arabic (c. 773) the Islamic scientific world began to be acquainted with the so-called Hindu system. This system began to be more widely used in the Arabic world and beyond, though the Greek system of numeration also remained in use as well as other native systems. Social factors may have played a role—the Oriental tradition favoring the decimal place-value method against the method of the Greeks. The symbols used to express the place-value numerals show wide variations but there are two main types: the Hindu symbols used by the Eastern Arabs, and the so-called gobâr (or ghubār)

---

[2]C. T. Rajagopal and T. V. Vedaminthi Aiyar, Scripta math., Vol. 17 (1951), pp. 65–74; Ibid., Vol. 18 (1952), pp. 25–30. See also C. T. Rajagopal and A. Venkataraman, J. Roy. Asiatic Soc. Bengali., Vol. 15, No. 1 (1949), pp. 1–13; and J. E. Hofmann, Math. Phys. Seminar Ber. (Giessen), Vol. 3 (1953), pp. 193–206.

[3]This may be compared to the use of the concept of the "void" (kenos) in Aristotle's Physica IV. 8.215ᵇ. See also C. B. Boyer, "Zero: the Symbol, the Concept, the Number," Nat. Math. Mag., Vol. 18 (1944), pp. 323–30.

[4]Cf. H. Freudenthal, 5000 jaren internationale wetenschap (Groningen, 1946).

numerals used in Spain among the Western Arabs. The first symbols are still used in the Arab world but our present numeral system seems to be derived from the *gobâr* system. There exists the theory of Woepcke (see Chap. 3, Sec. 11, above), according to which the *gobâr* numerals were in use in Spain when the Arabs arrived, having reached the West through the Neo-Pythagoreans of Alexandria as early as the year A.D. 450.[5]

4

Mesopotamia, which under Hellenistic and Roman rule had become an outpost of the Roman empire, reconquered its central position along the trade routes under the Sassanians, who reigned as native Persian kings over Persia in the tradition of Cyrus and Xerxes. Little is known about many aspects of this period in Persian history, especially about its science, but the legendary history—as reflected in the *Thousand and One Nights*, Omar Khayyam, Firdawsî—confirms the meager historical record that the Sassanian period was an era of cultural splendor. Situated between Constantinople, Alexandria, India, and China, Sassanian Persia was a country where many cultures met. Babylon had disappeared but was replaced by Seleucia-Ctesiphon, which was in turn replaced by Bagdad after the Arabic conquest of 641. This conquest left much of ancient Persia unaffected, though Arabic replaced Pehlevi as the official language. Even Islam was only accepted in a modified form (Shî'ism); Christians, Jews, and Zoroastrians continued to contribute to the cultural life of the Bagdad caliphate.

The mathematics of the Islamic period shows the same blend of influences with which we have become familiar in Alexandria and in India.[6] The Abbâsid caliphs, notably Al-Mansûr (754–775), Hârûn al-Rashîd (786–809), and Al-Ma'mûn (813–833), promoted astronomy and mathematics, Al-Ma'mûn even organizing at Bagdad a "House of Wisdom" with a library and an observatory. Islamic activities in the exact sciences, which began with Al-Fâzarî's translation of the *Siddhântâs*, reached its first height with a native from Khiva, Muhammad ibn Mûsâ al-Khwârizmî, who flourished about 825. Muhammad wrote several books on mathematics and astronomy. His arithmetic explained the Hindu system of numeration. Although this book is lost in the original Arabic, a Latin translation of the twelfth century is extant. It was one of the means by which Western Europe became acquainted with the decimal position. The title of the translation, *Algorithmi de numero Indorum*, added the term *algorithmus*—a latinization of the author's name—to our mathematical language. Something similar happened to Muhammad's algebra, which had the title *Hisâb al-jabr wal-muqâbala* (literally, "science of reduction and confrontation," probably meaning "science of equa-

---

[5]Cf. S. Gandz, "The Origin of the Ghubâr Numerals," *Isis*, Vol. 16 (1931), pp. 393–424. There exists also a theory of N. Bubnov, which holds that the *gobâr* forms were derived from ancient Greco-Roman symbols used on the abacus. See also the footnote in F. Cajori, *History of Mathematics* (New York, 1938), p. 90, as well as D. E. Smith and L. C. Karpinski, *The Hindu-Arabic Numerals* (Boston, 1911), p. 71.

[6]Accounts of medieval Oriental mathematics have been handicapped for a long time by the fact that so little source material was available in translation. Gradually the situation has improved, although some important contributions are still only available in Russian.

tions"). This algebra, of which the Arabic text is extant, also became known in the West through Latin translations, and they made the word *al-jabr* synonymous with the whole science of "algebra," which, indeed, until the middle of the nineteenth century was nothing but the science of equations.

This "algebra" contains a discussion of linear and quadratic equations, but without any algebraic formalism. Even the Diophantine "syncopated" symbolism is absent. Among these equations are the three types characterized by $x^2 + 10x = 39$, $x^2 + 21 = 10x$, $3x + 4 = x^2$, which had to be separately treated as long as positive coefficients were the only ones which were admitted. These three types reappear frequently in later texts—"thus the equation $x^2 + 10x = 39$ runs like a thread of gold through the algebras for several centuries," writes Professor L. C. Karpinski.[7] Much of the reasoning is geometric. Muhammad's astronomical and trigonometrical tables (with sines and tangents) also belong to the Arabic works which later were translated into Latin. His geometry is a simple catalogue of mensuration rules; it is of some importance because it can be directly traced to a Jewish text of the year A.D. 150. It shows a definite lack of sympathy with the Euclidean tradition. Al-Khwārizmī's astronomy was an abstract of the *Siddhāntās*, and therefore may show some indirect Greek influence by way of a Sanskrit text. The works of Al-Khwārizmī as a whole seem to show Oriental rather than Greek influence,[8] and this may have been deliberate.

Al-Khwārizmī's work plays an important role in the history of mathematics, for it is one of the main sources through which Indian numerals and Arabic algebra came to Western Europe. Algebra, until the middle of the nineteenth century, revealed its Oriental origin by its lack of an axiomatic foundation, in this respect sharply contrasting with Euclidean geometry. The present-day school algebra and geometry still preserve these tokens of their different origin.

5

The Greek tradition was cultivated by a school of Arabic scholars[9] who faithfully translated the Greek classics into Arabic—Apollonius, Archimedes, Euclid, Ptolemy, and others. The general acceptance of the name *Almagest* for Ptolemy's *Great Collection* shows the influence of the Arabic translations upon the West. This copying and translating has preserved many a Greek classic which otherwise would have been lost. There was a natural tendency to stress the computational and practical side of Greek mathematics at the cost of its theoretical side. Arabic astronomy was particularly interested in trigonometry—the word *sinus* is a Latin translation of the Arabic spelling of the Sanskrit *jyā*. The sines correspond to half the chord of the double arc (Ptolemy used the whole chord), and were conceived

---

[7]L. C. Karpinski, *Robert of Chester's Latin Translation of the Algebra of Al-Khwārismī* (New York, 1915), p. 19.

[8]S. Gandz, "The Sources of Al-Khwārizmī's Algebra," *Osiris*, Vol. 1 (1936), pp. 263–77.

[9]When we say "Arabic scholars," or "Arabic science" we do not mean that this science was only cultivated by the Arabs. On the contrary, most "Arabic scholars" were Persians, Tadjiks, Egyptians, Jews, Moors, etc. In the same way we can call many European writers from Boethius to Gauss "Latin scholars," because they wrote in Latin. Arabic was the *lingua franca* of the Islamic world, as Latin of the Western, and Greek of the Eastern Christian world.

as lines, not as numbers. We find a good deal of trigonometry in the works of Al-Battānī (Albategnius, c. 850–929), one of the great Arabic astronomers, who had a table of cotangents for every degree (umbra extensa) as well as the cosine rule for the spherical triangle.

This work by Al-Battānī shows that Arabic writers not only copied but also contributed new results through their mastery of both Greek and Oriental methods. Abū-l-Wafā (al-Būzjanī, 940–997/8) derived the sine theorem of spherical trigonometry, computed sine tables for intervals of 15′ of which values were correct in eight decimal places, introduced the equivalents of secant and cosecant, and played with geometrical constructions using a compass of one fixed opening. He also continued the Greek study of cubic and biquadratic equations. Al-Karkhī (or Al-Karajī; d. c. 1029), who wrote an elaborate algebra following Diophantus, had interesting material on surds, such as the formulas $\sqrt{8} + \sqrt{18} = \sqrt{50}$, $\sqrt[3]{54} - \sqrt[3]{2} = \sqrt[3]{16}$. He showed a definite tendency in favor of the Greeks; his "neglect of Hindu mathematics was such that it must have been systematic."[10]

6

We need not follow the many political and ethnological changes in the world of Islam. They brought ups and downs in the cultivation of astronomy and mathematics; certain centers disappeared, others flourished for a while; but the general character of the Islamic type of science remained virtually unchanged. We shall mention only a few highlights.

About the year A.D. 1000 new rulers appeared in Northern Persia, the Saljūq (Selchuk) Turks, whose empire flourished around the irrigation center of Merv. Here lived Omar Khayyam (or al-Khayyāmī; c. 1038/48–1123/31), known in the West as the author of the Rubaiyat (in the free FitzGerald version, 1859); he was an astronomer and a philosopher (in the spirit of Aristotle):

> Ah, but my Computations, People say,
> Have squared the Year to human Compass, eh?
> If so, by striking from the Calendar
> Unborn To-morrow, and dead Yesterday.

(LIX)

Here Omar may have referred to his reform of the old Persian calendar, which instituted an error of one day in 5000 years (1540 or 3770 years, according to other interpretations), whereas our present Gregorian calendar has an error of one day in 3330 years. His reform was introduced in 1079, but was later replaced by the Muslim lunar calendar. Omar wrote an Algebra,[11] which represented a considerable achievement since it contained a systematic investigation of cubic equations. Employing a method occasionally used by the Greeks, he determined

---

[10]G. Sarton, Introduction to the History of Science (Baltimore, 1927), Vol. I, p. 719.

[11]Risala fi'l-barāhīn 'alā masā'il al-jabr wa'l muqābala (Treatise on Demonstration of Problems of Reduction and Confrontation [meaning problems of equations]). See DSB, Vol. 7 (1973), p. 327.

TOMB OF OMAR KHAYYAM IN NISHAPUR.
Courtesy of the Metropolitan Museum of Art.

the roots of these equations as the intersection of two conic sections. He had no numerical solutions and discriminated—also in Greek style—between "geometrical" and "arithmetical" solutions, the latter existing only if the roots are positive rational. This approach was therefore entirely different from that of the sixteenth-century Bolognese mathematicians, who used purely algebraical methods.

Omar, in another book dealing with difficulties in Euclid, replaced the parallel axiom by a set of other propositions. Here he came to the figures now associated with the so-called "hypotheses of the obtuse, acute, and right angle," which now find their place in non-Euclidean geometry. He also replaced Euclid's theory of proportions by a numerical theory, in which he approached a theory of the irrational, and the concept of the real number in general.[12]

After the sack of Bagdad in 1258 by the Mongols, a new center of learning sprang up near the same place at the observatory of Maragha, built by the Mongol ruler Hulagu for Naṣīr al-dīn at Ṭūsī (Nasir-eddin, 1201–1274). Here again rose an institute where the whole of Oriental science could be pooled and matched with the Greek. Naṣīr separated trigonometry as a special science from astronomy; his attempts to "prove" Euclid's parallel axiom following in the path of Omar Khayyam show that he appreciated the theoretical approach of the Greeks. Naṣīr's influence was widely felt later in Renaissance Europe; as late as 1651 and 1663 John Wallis used Naṣīr's work on the Euclidean postulate. Naṣīr also continued in Omar's tradition in his theory of ratio and the new numerical approach to the irrational.

Another Persian mathematician, Jemshid Al-Kāshī (d. *c.* 1436), showed a great versatility in numerical work, comparable to that later reached by late sixteenth-century Europeans. He solved cubic equations by iteration and by trigonometric methods, and knew the method now known as Horner's for the solution of general algebraic equations of higher order which generalizes the extraction of roots of higher order or ordinary numbers—a method which seems to point to Chinese influence. In his work we find the binomial formula for a general positive integer exponent.[13] Next to sexagesimal fractions he uses decimal fractions with ease (e.g., 25.07 times 14.3 is 357.501) and has π in 16 decimals.[14] This all may indicate that the Chinese mathematics of the Song (Sung) dynasty had been penetrating deeply into the Islamic world (see Section 7, following).

An important figure in Egypt was Ibn Al-Haitham (Alhazen, *c.* 965–1039), the greatest Muslim physicist, whose *Optics* had a great influence on the West. He solved the "problem of Alhazen," in which we are asked to draw from two points on the plane of a circle lines meeting at the point of the circumference and making equal angles with the normal at that point. This problem leads to a biquadratic equation and was solved in the Greek way by a hyperbola meeting a circle.

---

[12]D. J. Struik, "Omar Khayyam, Mathematician," *Mathematics Teacher*, Vol. 51 (1958), pp. 280–85.

[13]Cf. M. Yadagari, "The Binomial Theorem: A Widespread Concept in Medieval Islamic Tradition," *HM*, Vol. 7 (1980), pp. 401–06.

[14]We already find decimal fractions in a book by Al-Uglīdīsī, who wrote it in Damascus in 952/953. See "Literature," below.

Alhazen also used the exhaustion method to compute the volumes of figures obtained by revolving a parabola about any diameter or ordinate. One hundred years before Alhazen there lived in Egypt the algebraist Abū Kāmil, who followed and extended the work of Al-Khwārizmī. He influenced not only Al-Karkhī, but also Leonardo of Pisa.

Another center of learning existed in Spain. One of the important astronomers at Cordoba was Al-Zarqāli, (Arzachel, *c.* 1029–1087), the best observer of his time and the editor of the so-called Toledan planetary tables. The trigonometrical tables of this work, which was translated into Latin, had some influence on the development of trigonometry in the Renaissance. The Toledan tables themselves were followed by the Alfonsine tables (after Alfonso X of Castile, 13th century), which were for centuries authoritative.

7

As to Chinese mathematics, it would be wrong to consider it as an isolated phenomenon, perhaps like the mathematics of the Maya. There always existed, at the very least from the time of the Han (contemporary with the Roman empire), considerable commercial and cultural relations with other parts of Asia and even Europe. Indian and later Arabic science had their influence on China, and Chinese science in turn left its imprint on the science of other countries. We think, for instance, of the decimal position system and negative numbers, which may well have found their way from China into India. Indian influence on China may again be as old as the introduction of Buddhism into China (first century A.D.). However, there is little or no evidence of Greek influence despite some parallel developments.

The studies on the ratio of circumference and diameter of the circle, which are typical of the post-Han period, seem therefore to have been conducted independently of Archimedes. Liu Hui, author of an extant commentary on the *Nine Chapters* (A.D. 263), found by inscribed and circumscribed regular polygons that $3.1401 < \pi < 3.1427$, and two centuries later we find with Zu Chongzhi (Tsu Ch'ung-chih; 430–501) and his son not only a value of π in seven decimals, but also the values $\pi = \frac{22}{7}$ and $\frac{355}{113}$.[15]

During the Tang (T'ang) dynasty (618–906) a corpus of the most important mathematical books was assembled for official use in the imperial examinations. It was in this period that printing began, although the first printed mathematical books of which we have knowledge date from 1084 and later. In 1115 a printed edition of the *Nine Chapters* appeared.

In a book written by Wang Xiaotong (Wang Hsiao-t'ung) around 625 we find a cubic equation more complicated than the $x^3 = a$ of the *Nine Chapters*. But the

---

[15] $\pi = \dfrac{355}{113}$ can be obtained from the values of Ptolemy and Archimedes: $= \dfrac{355}{113} = \dfrac{377 - 22}{120 - 7}$.

This value, which appears as a partial fraction when π, written in decimals, is expanded in a continuous fraction, is sometimes referred to as the "value of Metius," after the Alkmaar burgomaster and military engineer Adriaen Anthoniszoon (*c.* 1580), whose sons called themselves Metius.

period of greatest flowering of ancient Chinese mathematics was that of the Song (Sung) dynasty (960–1279), and the early years of the Mongol dynasty of the Yuan (the "Great Khan" of Marco Polo fame). Among the leading figures we find Qin Jiushao (Ch'in Chiu-shao), whose book dates from 1247, and who developed the theory of indeterminate equations, a theory which had been gradually developed throughout the centuries. One of his examples can be written

$$x \equiv 32 \ (\text{mod } 83) \equiv 70 \ (\text{mod } 110) \equiv (30 \ \text{mod } 135).$$

Qin also dealt with the numerical solution of higher-degree equations, as exemplified by

$$-x^4 + 763200x^2 - 40642560000 = 0.$$

He solved such equations by generalizing the method of successive approximations, used already in the *Nine Chapters*, for the extraction of square and cubic roots. In this method we now recognize the same operations known in our textbooks after W. G. Horner, who published it in 1819, apparently unaware that he had hit upon a method typical of a thousand years of Chinese mathematics.

Another mathematician of the Song period is Yang Hui, who worked with decimal fractions and wrote them in a way which reminds us of our present methods (book of 1261). One of his problems leads to $24.68 \times 36.56 = 902.3108$. Yang Hui presents us with the earliest extant representation of the Pascal triangle,[16] which we find again in a book *c.* 1303 written by Zhu Shijie (Chu Shih-chieh). Zhu, considered the most important of this group of mathematicians, offers in his books the most accomplished presentation of the Chinese arithmetical-algebraic-computational method. He even extended the "matrix" solution of systems of linear algebraic equations to equations of higher degree with several unknowns with methods that remind us of Sylvester in the nineteenth century.

In the post-Song period mathematical activity continued, but did not reach new heights. Speaking generally we can say that in their ability to do complicated arithmetical and algebraic work the Chinese mathematicians were not only comparable to their Indian colleagues and to those who wrote in Arabic, but that they occasionally served them as masters. For instance, the method of Horner and the work with decimal fractions can later be found in the books of Al-Kāshī of Samarkand (see Sec. 6, above).[17]

---

[16]L. Y. Lam, "The Chinese Connection between the Pascal Triangle and the Solution of Numerical Equations of Any Degree," *HM*, Vol. 7 (1980), pp. 407–24.

[17]By the sixteenth and seventeenth centuries Chinese and Japanese mathematicians were in contact with Europe. Western mathematics and astronomy were introduced into China by Father Matteo Ricci, who stayed in Peking from 1583 until his death in 1610. See H. Bosmans, "L'oeuvre scientifique de Mathieu Ricci, S. J.," *Revue des Questions Scient.*, 3rd ser., Vol. 29 (Jan., 1921), pp. 135–51.

Literature

The reader is referred to the references following Chapter II, in addition to the following.

Juschkewitsch (Juškevič), A. P. *Geschichte der Mathematik im Mittelalter*. Leipzig, 1964. (Original in Russian, Moscow, 1961).

[Vogel, K.] *Mohammed ibn Musa Alchwarizmi's Algorismus. Das früheste Lehrbuch zum Rechnen mit indischen Ziffern*. Aalen, 1963. (See S. Gandz, in *Quellen und Studien*, Vol. 2A (1932), pp. 61–85.)

Suter, H. *Die Mathematiker und Astronomer der Araber und ihre Werke*. Leipzig, 1900 (Suppl., 1902). (Also see H. P. J. Renaud, *Isis*, Vol. 18 [1932], p. 126.)

Kasir, D. S. *The Algebra of Omar Khayyam*. New York, 1931, p. 126.

Khayyam, Omar. *Traktaty*. Moscow, 1961 (in Russian). (Trans. by B. A. Rozenfel'd, with photostats of the Arabic texts.)

On Khayyam, see also:

Amir-Moéz, A. R. In *Scripta math.*, Vol. 26 (1963), pp. 323–37; also in *Istor.-mat. Issled.*, Vol. 15 (1963), pp. 445–72, 769–72.

Youschkevitch (Juškevič), A. P. *Les mathématiciens arabes (VIIIᵉ–XVᵉ siècle)*. Paris, 1976. Based on the author's Russian text; see above.

Rashad, R. "Résolution des équations numériques et algèbre: Sarafal Dinal-Tusi, Viète." *AHES*, Vol. 12 (1974), pp. 244–90.

Al-Kāshī, J. *The Key to Arithmetic, Treatise on the Circle*. Moscow, 1956 (in Russian). (Trans. by B. A. Rozenfel'd.)

Luckey, P. *Die Rechenkunst bei Ğamšid b. Mas'ūd al-Kāsī*. Wiesbaden, 1951. Also in *Abh. Deut. Akad. Wiss. Berlin, Klasse für Math. 1950*, No. 6 (1953).

——. "Die Ausziehung der n-ten Wurzel und der binomische Lehrsatz in der islamischen Mathematik." *Math. Annalen*, Vol. 20 (1948), pp. 217–74.

Juschkewitsch (Juškevič), A. P., and Rozenfel'd, B. A. "Die Mathematik der Länder des Ostens im Mittelalter." *Sowjetische Beiträge zur Geschichte der Naturwissenschaft*, Berlin (1960), pp. 62–160.

Saidan, A. S. *The Arithmetic of Al-Uglīdīsī*. Dordrecht, 1978.

Hankel, H. *Zur Geschichte der Mathematik im Altertum und Mittelalter*. Leipzig, 1874. (This book, the first to give a modern introduction to Arabic mathematics, was reedited by J. E. Hofmann, Hildersheim, 1965.)

Wang, Ling, and Needham, J. "Horner's Method in Chinese Mathematics." *T'oung Pao*, Leiden, Vol. 43 (1955), pp. 345–401.

Datta, B. *The Science of the Sulba, a Study in Early Hindu Geometry*. London, 1932.

Smith, D. E., and Karpinski, L. C. *The Hindu-Arabic Numerals*. Boston, 1911.

Karpinski, L. C. *Robert of Chester's Latin Translation of the Algebra of Al-Khwārizmī*. New York, 1915.

Hayashi, T. "A Brief History of the Japanese Mathematics." *Nieuw Archief voor Wiskunde*, 2nd ser., Vol. 6 (1904–05), pp. 296–361.

Smith, D. E. "Unsettled Questions Concerning the Mathematics of China." *Scientific Monthly*, Vol. 33 (1931), pp. 244–50.

Lam, L. Y. *A Critical Study of the Yang Hui Suan Fa: A Thirteenth Century Chinese Mathematical Treatise.* Singapore, 1977. (Cf. J. Needham, *HM*, Vol. 6 (1979), pp. 466–68).

Colebrooke, H. T. *Algebra, with Arithmetic and Mensuration from the Sanskrit of Brahmagupta and Bhāscara.* London, 1817. 2nd ed. (revised by H. C. Banerji), Calcutta, 1927.

Srinivasiengar, C. N. *The History of Ancient Indian Mathematics.* Calcutta, 1967.

Kūshyār ibn Labbān. *Principles of Hindu Reckoning.* Trans. and ed. by M. Levey and M. Petruck. Madison, Wis., 1965.

Pingree, D. *History of Mathematical Astronomy in India. DSB*, Vol. 15, Supplement 1 (1978), pp. 533–633. (By the same author, see *DSB* articles on ancient Indian mathematicians such as Aryabhata.)

Clark, W. E. *The Aryabhatya of Aryabhata.* Chicago, 1930.

CHAPTER V

The Beginnings in Western Europe

1

The most advanced section of the Roman Empire from both an economic and a cultural point of view had always been the East. The western part had never been based on an irrigation economy; its agriculture was of the extensive kind which did not stimulate the study of astronomy. Actually the West managed very well in its own way with a minimum of astronomy, some practical arithmetic, and some mensuration for commerce and surveying; but the stimulus to promote these sciences came from the East. When East and West separated politically this stimulation almost disappeared. The static civilization of the Western Roman Empire continued with little interruption or variation for many centuries; the Mediterranean unity of antique civilization also remained unchanged—and was not even very much affected by the barbaric conquests. In all Germanic kingdoms, except perhaps those of Britain, the economic conditions, the social institutions, and the intellectual life remained fundamentally what they had been in the declining Roman Empire. The basis of economic life was agriculture, with slaves gradually replaced by free and tenant farmers; but in addition there were prosperous cities and a large-scale commerce with a money economy. The central authority in the Greco-Roman world after the fall of the Western Empire in 476 was shared by the emperor in Constantinople and the popes of Rome. The Catholic Church of the West through its institutions and language continued as best it could the cultural tradition of the Roman Empire among Germanic kingdoms. Monasteries and cultured laymen kept some of the Greco-Roman civilization alive.

One of these laymen, the diplomat and philosopher Anicius Manlius Severinus Boethius, wrote mathematical texts which were considered authoritative in the Western world for more than a thousand years. They reflect the cultural conditions, for they are poor in content and their very survival may have been influenced by the belief that the author died in 524 as a martyr of the Catholic faith. His *Institutio arithmetica*, a superficial translation of Nicomachus, did provide some Pythagorean number theory which was absorbed in medieval institutions as part of the seven *artes liberales*, consisting of a *trivium* (grammatica, rhetorica, dialectica) and a *quadrivium* (arithmetica, geometrica, astronomia, musica), terms that were used first, it seems, by Boethius.

It is difficult to establish the period in the West in which the economy of the ancient Roman Empire disappeared to make room for a new feudal order. Some light on this question may be shed by the hypothesis of H. Pirenne,[1] according to which the end of the ancient Western world came with the expansion of Islam. The Arabs dispossessed the Byzantine Empire of all its provinces on the eastern and southern shores of the Mediterranean and made the Eastern Mediterranean a closed Muslim lake. They made commercial relations between the Near Orient and the Christian Occident extremely difficult for several centuries. The intellectual avenue between the Arabic world and the northern parts of the former Roman Empire, though never wholly closed, was obstructed for centuries.

Then in Frankish Gaul and other former parts of the Roman Empire large-scale economy subsequently vanished; decadence overtook the cities; returns from tolls became insignificant. The money economy was replaced by barter and local marketing. Western Europe, in short, was reduced to a state of semibarbarism. The landed aristocracy rose in significance with the decline of commerce; the North Frankish landlords, headed by the Carolingians, became the ruling power in the land of the Franks. The economic and cultural center moved to Northern France and Britain. The separation of East and West limited the effective authority of the Pope to the extent that the papacy allied itself with the Carolingians, a move symbolized by the crowning of Charlemagne in 800 as Emperor of the Holy Roman Empire. Western society became feudal and ecclesiastical, its orientation Northern and Germanic. It broke with the static tradition inherited from the Roman Empire in its later centuries. The novel harnessing of horses and the introduction of the stirrup are but a few examples of this break. When cities developed, the class of burghers was open to this inventiveness.

2

During the early centuries of Western feudalism, however, we find little appreciation of mathematics even in the monasteries. In the again primitive agricultural society of this period the factors stimulating mathematics, even of a directly practical kind, were nearly nonexistent; and monastic mathematics was no more than some ecclesiastical arithmetic used mainly for the computation of Eastertime (the so-called *computus*). Boethius was the highest source of authority. Of some importance among these ecclesiastical mathematicians was the British-born Alcuin, associated with the court of Charlemagne, whose *Problems for the Quickening of the Mind of the Young* (see Chap. IV, Sec. 3, above) contained a selection of problems which have influenced the writers of textbooks for many centuries. Many of these problems date back to the ancient Orient. For example:

A dog chasing a rabbit, which has a start of 150 feet, jumps 9 feet every time the rabbit jumps 7. In how many leaps does the dog overtake the rabbit?

---

[1]H. Pirenne, *Mahomet et Charlemagne* (Paris, 1937; English translation, New York, 1939). The thesis has been attacked, especially by A. Doepsch. The transition, it is claimed, was rather due to internal conditions. See A. F. Havighurst, *The Pirenne Thesis* (Boston, 1958).

A wolf, a goat, and a cabbage must be moved across a river in a boat holding only one beside the ferry man. How must he carry them across so that the goat shall not eat the cabbage, nor the wolf the goat?

Another ecclesiastical mathematician was Gerbert, a French monk, who in 999 became Pope under the name of Sylvester II. He wrote some treatises in the spirit of Boethius and was active in awakening interest in mathematics throughout Western Europe. An abacus with a board of as many as twenty-seven columns has been ascribed to Gerbert or his influence. His stay in Catalonia around 968 may have contributed to the willingness of the Latin world to study Arabic science.

3

There are significant differences between the development of Western, early Greek, and Oriental feudalism. The extensive character of Western agriculture made a vast system of bureaucratic administrators superfluous, so that it could not supply a basis for an eventual Oriental despotism. There was no possibility in the West of obtaining vast supplies of slaves. When villages in Western Europe grew into towns these towns developed into self-governing units, in which the burghers were unable to establish a life of leisure based on slavery. This is one of the main reasons why the development of the Greek *polis* and that of the Western city, which during the early stages had much in common, deviated sharply in later periods. The medieval townspeople had to rely on their own inventive genius to improve their standard of living. Fighting a bitter struggle against the feudal landlords—and with much civil strife in addition—they emerged victorious in the twelfth, thirteenth, and fourteenth centuries. This triumph was based not only on a rapid expansion of trade and money economy, but also on a gradual improvement in technology. The feudal princes often supported the cities in their fight against the smaller landlords, and then eventually extended their rule over the cities. This finally led to the emergence of the first national states in Western Europe.

The cities began to establish commercial relations with the Orient, which was still the center of civilization. Sometimes these relations were established in a peaceful way, sometimes by violent means as in the many Crusades. First to establish mercantile relations were the Italian cities; they were followed by those of France and Central Europe. Scholars followed, or sometimes preceded, the merchant and the soldier. Spain and Sicily were the nearest points of contact between East and West, and here Western merchants and students became acquainted with Islamic civilization. When in 1085 Toledo was taken from the Moors by the Christians, Western students flocked to this city to learn science as it was transmitted in Arabic. They often employed Jewish interpreters to converse and to translate, and so we find in twelfth-century Spain, Plato of Tivoli, Gherardo of Cremona, Adelard of Bath, and Robert of Chester, translating Arabic mathematical manuscripts into Latin. Thus Europe became familiar with Greek classics through the Arabic; and by this time Western Europe was advanced enough to appreciate this knowledge.

Another center of learning was Constantinople (now Istanbul), after 395 for more than a thousand years the capital of the East Roman Empire. Here the Greek heritage was preserved as well as possible, and gave opportunity to Latin scholars to study the Greek classics directly, without recourse to Arabic (or Syriac or Hebrew) translations.

4

As we have said, the first powerful commercial cities arose in Italy, where during the twelfth and thirteenth centuries Genoa, Pisa, Venice, Milan, and Florence carried on a flourishing trade between the Arabic world and the North. Italian merchants visited the Orient and studied its civilization; Marco Polo's travels show the intrepidity of these adventurers. Like the Ionian merchants of almost two thousand years before, they tried to study the science and the arts of the older civilization not only to reproduce them, but also to assimilate them into their own mercantile society, which as early as in the twelfth and thirteenth centuries saw the growth of banking and the beginnings of a capitalist form of industry. The first Western merchant whose mathematical studies showed considerable maturity was Leonardo of Pisa.

Leonardo, also called Fibonacci ("member of the house of the Bonacci"), traveled in the Orient as a merchant. On his return he wrote his *Liber Abaci* (1202), filled with arithmetical and algebraical information which he had collected on his travels. In the *Practica Geometriae* (1220), Leonardo described in a similar way whatever he had discovered in geometry and trigonometry. He may have been an original investigator as well, since his books contain many examples which seem to have no exact duplicates in Arabic literature.[2] However, he does quote Al-Khwārizmī, as for instance in the discussion of the equation $x^2 + 10x = 39$. The problem which leads to the "series of Fibonacci"—0, 1, 1, 2, 3, 5, 8, 13, 21, . . .—of which each term is the sum of the two preceding terms, seems to be new, and also his remarkably mature proof that the roots of the equation $x^3 + 2x^2 + 10x = 20$ cannot be expressed by means of Euclidean irrationalities $\sqrt{a} \pm \sqrt{b}$ (hence cannot be constructed by means of compasses and ruler only). Leonardo proved it by checking upon each of Euclid's fifteen cases, and then solved the positive root of this equation approximately, finding six sexagesimal places.

The series of Fibonacci resulted from the problem:

How many pairs of rabbits can be produced from a single pair in a year if (*a*) each pair begets a new pair every month, which from the second month on becomes productive, and (*b*) deaths do not occur?

The *Liber Abaci* is one of the means by which the Hindu-Arabic system of numeration was introduced into Western Europe. Its occasional use dates back some centuries before Leonardo, when it was imported by merchants, ambassadors, scholars, pilgrims, and soldiers coming from Spain and from the Levant. The oldest dated European manuscript containing the numerals is the *Codex*

---

[2]L. C. Karpinski (*Amer. Math. Monthly*, Vol. 21 [1914], pp. 37–48), using the Paris manuscript of Abū Kāmil's algebra, claims that Leonardo followed Abū Kāmil in a whole series of problems.

Vigilanus, written in Spain in 976. However, the introduction of the ten symbols into Western Europe was slow; the earliest French manuscript in which they are found dates from 1275. The Greek system of numeration remained in vogue along the Adriatic for many centuries. Computation was often performed on the ancient abacus, a board with counters or pebbles (often simply consisting of lines drawn in sand) similar in principle to the counting boards still used by the Russians, Chinese, and Japanese, and by children on their playpens. Roman numerals were used to registrate the result of a computation on the abacus. Throughout the Middle Ages (and even later) we find Roman numerals in merchants' ledgers, which indicates that the abacus was used in the offices. The introduction of Hindu-Arabic numerals met with opposition from the public, since the use of these symbols made merchants' books difficult to read. In the statutes of the *Arte del Cambio* of 1299 and even later the money changers of Florence were forbidden to use Arabic numerals and were obliged to use Roman ones. Sometime during the fourteenth century Italian merchants began to use some Arabic figures in their account books.[3] Occasionally we find intermediate forms such as II^{m}III^{c}XV for 2315.

5

With the extension of trade, interest in mathematics spread slowly to the northern cities. It was at first mainly a practical interest, and for several centuries arithmetic and algebra were taught outside the universities by self-made *Rechenmeister* ("reckon masters," arithmeticians), usually ignorant of the classics, who also taught bookkeeping and navigation. For a long time this type of mathematics kept definite traces of its Arabic origin, as words such as "algebra" and "algorithm" testify.

Speculative mathematics did not entirely die during the Middle Ages, though it was cultivated not among the men of practice, but among the scholastic philosophers. Here the study of Plato and Aristotle, combined with meditations on the nature of the Deity, led to subtle speculations on the nature of motion, of the continuum, and of infinity. Origen had followed Aristotle in denying the existence of the actually infinite, but St. Augustine in the *Civitas Dei* had accepted the whole sequence of integers as an actual infinity. His words were so well chosen that Georg Cantor has remarked that the transfinitum cannot be more energetically desired and cannot be more perfectly determined and defended than was done by St. Augustine.[4] The scholastic writers of the Middle Ages, especially St.

---

[3]In the Medici account books (dating from 1406) of the Selfridge Collection on deposit at the Harvard Graduate School of Business Administration, Hindu-Arabic numerals frequently appear in the narrative or descriptive column. From 1439 onward they replace Roman numerals in the money or effective column of the books of primary entry—journals, wastebooks, etc.—but not until 1482 were Roman numerals abandoned in the money column of the business ledgers of all but one Medici merchant. From 1494, only Hindu-Arabic numerals are used in all the Medici account books. (From a letter by Dr. Florence Edler De Roover.) See also, F. Edler, *Glossary of Medieval Terms of Business* (Cambridge, Mass., 1934), p. 389. Also: D. J. Struik, "The prohibition of the use of Arabic numerals in Florence," *Arch. intern. hist. sc.*, Vol. 21 (1968), pp. 291–94.

[4]G. Cantor, "Letter to Eulenburg (1886)," *Ges. Abhandlugen* (Berlin, 1932), pp. 401–02. The passage quoted by Cantor, Ch. XVIII of Book XII of *The City of God* (in the Healey translation), is entitled "Against such as say that things infinite are above God's knowledge."

Thomas Aquinas, accepted Aristotle's *infinitum actu non datur*,[5] but considered every continuum as potentially divisible ad infinitum. Thus there was no smallest line. A point, therefore, was not a part of a line, because it was indivisible: *ex indivisibilibus non potest constare aliquod continuum.*[6] A point could generate a line by motion. Such speculations had their influence on the inventors of the infinitesimal calculus in the seventeenth century and on the philosophers of the transfinite in the nineteenth; Cavalieri, Tacquet, Bolzano, and Cantor knew the scholastic authors and pondered over the meaning of their ideas.

These churchmen occasionally reached results of more immediate mathematical interest. Thomas Bradwardine, who became Archbishop of Canterbury, investigated star polygons after studying Boethius. The most important of these medieval clerical mathematicians was Nicole Oresme, Bishop of Lisieux in Normandy, who played with fractional powers. Since $4^3 = 64 = 8^2$, he wrote 8 as

$$\boxed{1^p\tfrac{1}{2}}\ 4 \quad \text{or} \quad \boxed{\dfrac{p.1}{1.2}}\ 4, \quad \text{meaning } 4^{1\frac{1}{2}}.$$ He also wrote a tract called *De latitudinibus formarum* (*c.* 1360), in which he graphs a dependent variable (*latitudo*) against an independent one (*longitudo*), which is subjected to variation. It shows a kind of vague transition from coordinates on the terrestrial or celestial sphere, known to the Ancients, to modern coordinate geometry. This tract was printed several times between 1482 and 1515 and may have influenced Renaissance mathematicians, including Descartes. Oresme also wrote an infinite series, showing that the harmonic series is divergent, a remarkable result for his day.

6

The main line of mathematical advance passed through the growing mercantile cities under the direct influence of trade, navigation, astronomy, and surveying. The townspeople were interested in counting, in arithmetic, in computation. Sombart labeled this interest of the fifteenth- and sixteenth-century burgher his *Rechenhaftigkeit.*[7] Leaders in the love for practical mathematics were the "reckon masters," only very occasionally joined by a university man, able, through his study of astronomy, to understand the importance of improving computational methods. Centers of the new life were the Italian cities and the Central European cities of Nuremberg, Vienna, and Prague. The fall of Constantinople in 1453, which ended the Byzantine Empire, led many Greek scholars to the Western cities. Interest in the original Greek texts increased, and it became easier to satisfy this interest. University professors joined with cultured laymen in studying the texts, ambitious "reckon masters" listened and tried to understand the new knowledge in their own way.

---

[5]"There is no actually infinite." See further E. Bodewig, "Die Stellung des hl. Thomas von Aquino zur Mathematik," *Arch. f. Geschichte der Philosophie,* Vol. 41 (1932), pp. 408–34.

[6]"A continuum cannot consist of indivisibles."

[7]W. Sombart, *Der Bourgeois* (Munich and Leipzig, 1913), p. 164. The term *Rechenhaftigkeit* indicates a willingness to compute, a belief in the usefulness of arithmetical work.

Typical of this period was Johannes Müller of Königsberg,[8] or Regiomontanus, the leading mathematical figure of the fifteenth century. The activity of this remarkable instrument maker, printer, and scientist illustrates the advances made in European mathematics during the two centuries after Leonardo. He was active in translating and publishing the classical mathematical manuscripts available. His teacher, the Viennese astronomer, George Peurbach—author of astronomical and trigonometrical tables—had already begun a translation of the astronomy of Ptolemy from the Greek. Regiomontanus continued this translation and also translated Apollonius, Heron, and the most difficult of all, Archimedes. His main original work was *De triangulis omnimodis libri quinque* (1464, not printed until 1533), a complete introduction to trigonometry, differing from our present-day texts primarily in the fact that our convenient notation did not exist. It contains the law of sines in a spherical triangle. All theorems had still to be expressed in words. Trigonometry, from that point on, became a science independent of astronomy. Naṣīr al-dīn had accomplished something similar in the thirteenth century, but it is significant that his work never resulted in much further progress, whereas Regiomontanus' book deeply influenced further development of trigonometry and its application to astronomy and algebra. Regiomontanus also devoted much effort to the computation of trigonometric tables. He has, for instance, tables of sines to radius 60.000 for intervals of one minute, which were printed after his death.

Sines were line segments, defined as semichords subtending angles in a circle. Their numerical values therefore depended on the length of the radius. A large radius allowed great accuracy in the value of the sines, without the necessity of introducing sexagesimal (or decimal) fractions. The systematic use of radius 1, and hence the concept of sines, tangents, etc., as ratios (numbers) is due to Euler (1748), who also introduced our present notation.

7

So far no definite step had been taken beyond the ancient achievements of the Greeks and Arabs. The classics remained the *ne plus ultra* of science. It came therefore as an enormous and exhilarating surprise when Italian mathematicians of the early sixteenth century actually showed that it was possible to develop a new mathematical theory which the ancients and Arabs had missed. This theory, which led to the general algebraic solution of the cubic equation, was discovered by Scipio del Ferro and his pupils at the University of Bologna.

The Italian cities had continued to show proficiency in mathematics after the time of Leonardo. In the fifteenth century their "reckon masters" were well versed in arithmetical operations, including surds (without having any geometrical scruples), and their painters were good geometers. Vasari, in his *Lives of the Painters*[9] (1550, 1564–68), mentions the interest in solid geometry among many

---

[8]This is not the city on the Baltic (present Kaliningrad), but a small city in Bavaria south of the Main.

[9]*Le vite de' più eccellenti pittori, scultori e architettori* (1550, enlarged ed. 1564–68). Several English translations exist, among them a Penguin selection (1965).

quattrocento engineers and artists. One of their achievements was the study of linear perspective, beginning with Brunelleschi, Donatello, and Uccello, followed by such men as Masaccio, Alberti, and Piero della Francesca. The last one wrote a book on solid bodies in perspective. Leonardo da Vinci and Raphael also were students of perspective.[10]

With the invention of printing, books for the teaching of practical arithmetic and its commercial applications began to be published, which spread the art far and wide.[11] The most impressive book on mathematics of these early days of printing was written by the Franciscan Luca Pacioli, appearing in 1494 as *Summa de Arithmetica, Geometria, Proportioni et Proportionalita* (folio, 600 pages).[12] Written in Italian—and not a very pleasant Italian—it contained all that was known in that day of arithmetic, algebra, geometry, and trigonometry. By then the use of Hindu-Arabic numerals was well established, and the arithmetical notation did not greatly differ from ours. Pacioli ended his book with the remark that the solution of the equations that in our present notation can be written $x^3 + mx = n$, $x^3 + n = mx$ seemed as impossible in the present state of science as the quadrature of the circle.

At this point began the work of the mathematicians at the University of Bologna. This university, around the turn of the fifteenth century, was one of the largest and most famous in Europe. Its faculty of astronomy alone at one time had sixteen lectors. From all parts of Europe students flocked to listen to the lectures—and to the public disputations that also attracted the attention of large, sportively minded crowds. Among the students at one time or another were Pacioli, Albrecht Dürer, and Copernicus. Characteristic of the new age was the desire not only to absorb classical information but also to create new things, to penetrate beyond the boundaries set by the classics. The art of printing and the discovery of America were examples of such possibilities. Was it possible to create new mathematics? Greeks and Orientals had tried their ingenuity on the solution of the third-degree equation but had only solved some special cases numerically. The Bolognese mathematicians now tried to find the general numerical solution.

These cubic equations could all be reduced to three types:

$$x^3 + px = q, \ x^3 = px + q, \ x^3 + q = px,$$

where p and q were positive numbers. They were specially investigated by Professor Scipio del Ferro, who died in 1526. It may be taken on the authority of E. Bortolotti[13] that del Ferro actually solved all types. He never published his

[10]S. Y. Edgerton, Jr., *The Renaissance Rediscovery of Linear Perspective* (New York, 1975).

[11]Among the first printed mathematical books were a commercial arithmetic (Treviso, 1478) and a Latin edition of Euclid's *Elements* (Venice, Ratdolt, 1482).

[12]Pacioli also published a book on the golden section, *Divina Proportione* (1509). Its beautiful figures, including those of star polyhedra, have been ascribed to Leonardo da Vinci. He also published the first treatise on double-entry bookkeeping, as part of his *Summa*.

[13]E. Bortolotti, "L'algebra nella scuola matematica bolognese del secolo XVI," *Periodico di Matematica*, ser. 4, Vol. 5 (1925), pp. 147–84.

solutions and only told a few friends about them. Nevertheless, word of the discovery became known, and after Scipio's death a Venetian "reckon master," nicknamed Tartaglia ("The Stammerer"), rediscovered his methods (1535). He showed his results in a public demonstration, but again kept the method by which he had obtained them a secret. Finally he revealed his ideas to a learned Milanese doctor, Girolamo Cardano (Jerome Cardan in English), who had to swear that he would keep them a secret. But when Cardano in 1545 published his rather short but stately book on algebra, with the proud title of *Ars magna*, Tartaglia discovered to his disgust that the method was fully disclosed in the book, with due acknowledgment to the discoverer, but stolen just the same. A bitter debate ensued, with insults hurled both ways, in which Cardano was defended by a younger gentleman scholar, Ludovico Ferrari. Out of this war came some interesting documents, among them the *Quaesiti* of Tartaglia (1546) and the *Cartelli* of Ferrari (1547–48), from which the whole history of this spectacular discovery became public knowledge. The *Ars magna* remained authoritative for many years.

The solution is now known as the Cardan solution, which for the case $x^3 + px = q$ takes the form

$$x = \sqrt[3]{\sqrt{\frac{p^3}{27} + \frac{q^2}{4}} + \frac{q}{2}} - \sqrt[3]{\sqrt{\frac{p^3}{27} + \frac{q^2}{4}} + \frac{q}{2}}$$

We see that this solution introduced quantities of the form $\sqrt[3]{a \pm \sqrt{b}}$, different from the Euclidean $\sqrt{a \pm \sqrt{b}}$.

Cardano's *Ars magna* contained another brilliant discovery: Ferrari's method of reducing the solution of the general biquadratic equation to that of a cubic equation. Ferrari's equation was $x^4 + 6x^2 + 36 = 60x$, which he reduced to $y^3 + 15y^2 + 36y = 450$. Cardano also considered negative numbers, calling them "fictitious," but was unable to do anything with the so-called "irreducible case" of the cubic equation in which there are three real solutions appearing as the sum or difference of what we now call complex numbers.

This difficulty was solved by the last of the great sixteenth-century Bolognese mathematicians, Raffael Bombelli, whose *Algebra* appeared in 1572. In this book—and in a geometry written around 1550 which remained in manuscript—he introduced a consistent theory of imaginary complex numbers. He wrote $3i$ as $\sqrt{0 - 9}$ (literally: $R[0\ m.\ 9]$, R for radix, m for meno). This allowed Bombelli to treat the irreducible case by showing, for instance, that

$$\sqrt[3]{52 + \sqrt{0 - 2209}} = 4 + \sqrt{0 - 1}.$$

Bombelli's book was widely read; Leibniz selected it for the study of cubic equations, and Euler quotes Bombelli in his own *Algebra* in the chapter on biquadratic equations. Complex numbers, from that time on, lost some of their supernatural character, though full acceptance came only in the nineteenth century.

It is a curious fact that the first introduction of the imaginaries occurred in the theory of the cubic equations, in the case where it was clear that real solutions exist though in an unrecognizable form, and not in the theory of quadratic equations, where our present textbooks usually introduce them.

8

Algebra and computational arithmetic remained for many decades the favorite subject of mathematical experimentation. Stimulation no longer came only from the *Rechenhaftigkeit* of the mercantile bourgeoisie but also from the demands made on surveying and navigation by the leaders of the new national states. Engineers were needed for the erection of public works and for military constructions. Astronomy remained, as in all previous periods, an important domain for mathematical studies. It was the period of the great astronomical theories of Copernicus, Tycho Brahe, and Kepler. A new conception of the universe emerged.

Philosophical thought reflected the trends in scientific thinking; Plato, with his admiration for quantitative mathematical reasoning, gained ascendancy over Aristotle. Platonic influence is particularly evident in Kepler's work. Trigonometrical and astronomical tables appeared with increasing accuracy, especially in Germany. The tables of the Wittenberg professor and friend of Copernicus, G. J. Rheticus, finished in 1596 by his pupil Valentin Otho, contain the values of all six trigonometric values for every ten seconds to ten places. They are known as *Opus Palatinum*. The tables of Pitiscus (1613) went up to fifteen places. The technique of solving equations and the understanding of the nature of their roots also improved. The public challenge, made in 1593 by the Belgian mathematician Adriaen van Roomen, to solve the equation of the 45th degree,

$$x^{45} - 45x^{43} + 945x^{41} - 12300x^{39} + \ldots - 3795x^3 + 45x = A,$$

was characteristic of the times. Van Roomen proposed special cases, e.g.,

$A = \sqrt{2 + \sqrt{2 + \sqrt{2 + \sqrt{2}}}}$, which gives $x = \sqrt{2 - \sqrt{2 + \sqrt{2 + \sqrt{2 + \sqrt{3}}}}}$, which cases were suggested by consideration of regular polygons.

François Viète, a French lawyer attached to the court of Henry IV, solved van Roomen's problem by observing that the left-hand member was equivalent to the expression of $\sin \phi$ in terms of $\sin \phi/45$. The solution could therefore be found with the aid of tables. Viète found twenty-three solutions of the form $\sin (\phi/45 - n.8°)$, discarding negative roots. Viète also reduced Cardano's solution of the cubic equation to a trigonometric one, in which process the irreducible case lost its horrors by making the introduction of imaginaries unnecessary. This solution can now be found in textbooks of "higher algebra."[14]

Viète's main achievements were in the improvement of the theory of equations (e.g., *In artem analyticam isagoge*, 1591), where he was among the first to represent numbers by letters. The use of numerical coefficients, even in the

---

[14]E.g., W. S. Burnside and A. W. Patton, *The Theory of Equations* (1892).

FRANÇOIS VIÈTE (1540–1603)

JOHN NAPIER (1550–1617)

"syncopated" algebra of the Diophantine school, had impeded the general discussion of algebraic problems. The work of the sixteenth-century algebraists (the "Cossists," after the Italian word *cosa* for the unknown) was produced in a rather complicated notation. But in Viète's *logistica speciosa*, at least a general symbolism appeared, in which letters were used to express numerical coefficients, though A^2 was still written as "*A* quadratum." Here we also find the signs + and − in our present meaning, but not for the first time. These may well have first appeared, at any rate in print, in a German arithmetic by Johann Widmann in 1489. Viète's *speciosa* also differs from our algebra by Viète's insistence on the Greek principle of homogeneity, in which a product of two line segments was necessarily conceived as an area; line segments could therefore only be added to line segments, areas to areas, and volumes to volumes. There was even some doubt whether equations of degree higher than three actually had a meaning, since they could only be interpreted in four dimensions, a conception hard to understand in those days and for a long time afterward.

This was the period in which computational technique reached new heights, and at last began to surpass the achievements of the Islamic world. Viète improved on Archimedes and found π in nine decimals; shortly afterward π was computed in thirty-five decimals by Ludolph van Coolen, a fencing master at Delft who used inscribed and circumscribed regular polygons with more and more sides. Viète also expressed π as an infinite product (1593), in our notation:

$$\frac{2}{\pi} = \cos\frac{\pi}{4} \cos\frac{\pi}{8} \cos\frac{\pi}{16} \cos\frac{\pi}{32}\cdots.$$

The improvement in technique was a result of the improvement in notation. The new results show clearly that it is incorrect to say that men like Viète "merely" improved notation. Such a statement discards the profound relation between content and form. New results have often become possible only because of a new mode of writing. The introduction of Hindu-Arabic numerals is one example; Leibniz's notation for the calculus is another one. An adequate notation reflects reality better than a poor one, and as such appears endowed with a life of its own which in turn creates new life. Viète's improvement in notation was followed, a generation later, by Descartes's application of algebra to geometry, and by our present notation.

9

Engineers and arithmeticians were in particular demand in the new commercial states, especially France, England, and the Netherlands. Astronomy was practiced all over Europe. Although the Italian cities were no longer on the main road to the Orient after the discovery of the sea route to India, they still remained centers of importance. And so we find among the great mathematicians and computers of the early seventeenth century Simon Stevin, an engineer, Johann Kepler and Thomas Harriot, astronomers, and Adriaen Vlacq and Ezechiel de Decker, surveyors.

Stevin, a bookkeeper of Bruges, became an engineer in the army of Prince Maurice of Orange, who appreciated the way Stevin combined practical sense with theoretical understanding and originality. In *La disme* (1585) he introduced decimal fractions as part of a project to unify the whole system of measurements on a decimal base. It was one of the great improvements made possible by the general introduction of the Hindu-Arabic system of numeration.

The other great computational improvement was the invention of logarithms. Several mathematicians of the sixteenth century had been playing with the possibility of coordinating arithmetical and geometrical progressions, mainly in order to ease the work with the complicated trigonometrical tables. An important contribution toward this end was undertaken by a Scottish laird, John Napier (or Neper), who in 1614 published *Mirifici logarithmorum canonis descriptio*. His central idea was to construct two sequences of numbers so related that when one increases in arithmetical progression, the other decreases in a geometrical one. Since the product of two numbers in the second sequence has a simple relation to the sum of the corresponding numbers in the first, multiplication could be reduced to addition. With this system Napier could considerably facilitate computational work with sines. Napier's early attempt was rather clumsy, since his two sequences correspond according to the modern formula

$$y = ae^{-x/a} \text{ (or } x = \text{Nep. log } y)$$

in which $a = 10^7$.[15] When $x = x_1 + x_2$, we do not get $y = y_1 y_2$, but $y = y_1 y_2/a$. This system did not satisfy Napier himself, as he told his admirer Henry Briggs, a professor at Gresham College, London. They decided on the function $y = 10^x$, for which $x = x_1 + x_2$ actually yields $y = y_1 y_2$. Briggs, after Napier's death, carried out this suggestion and in 1624 published *Arithmetica logarithmica* which contained the "Briggian" logarithms in 14 places for the integers from 1 to 20,000 and from 90,000 to 100,000. The gap between 20,000 and 90,000 was filled by Ezechiel de Decker, a Dutch surveyor, who, assisted by Vlacq, published at Gouda in 1627 a complete table of logarithms. The new invention was immediately welcomed by the mathematicians and astronomers, and particularly by Kepler, who had had a long and painful experience with elaborate computations.

Our explanation of logarithms by exponentials is historically somewhat misleading, since the concept of an exponential function dates only from the later part of the seventeenth century. Napier had no notion of a base. Natural logarithms, based on the function $y = e^x$, appeared almost contemporaneously with the Briggian logarithms, but their fundamental importance was not recognized until the infinitesimal calculus was better understood.[16]

---

[15]Hence Nep. log $y = 10^7 (\ln 10^7 - \ln y) = 161180957 - 10^7 \ln y$; and Nep. log $1 = 161180957$; ln x stands for our natural logarithm. It was also Napier who modified Stevin's notation for decimal fractions by introducing our notation with the decimal point. Decimal fractions, we have seen, were already used by Chinese and Arabic mathematicians long before Stevin.

[16]E. Wright, a writer on navigation, published some natural logarithms in 1618, J. Speidell in 1619, but after this no tables of these logarithms were published until 1770. See F. Cajori, "History of the Exponential and Logarithmic Concepts," *Amer. Math. Monthly*, Vol. 20 (1913), 7 articles.

Literature

For the spread of Hindu-Arabic numerals in Europe, see:

Smith, D. E., and Karpinski, L. C. *The Hindu-Arabic Numerals*. Boston and London, 1911.

For speculative mathematics in the Middle Ages, see:

Boyer, C. B. *The History of the Calculus*. New York, 1949. Dover reprint, 1959. (See Chapter III.)

Sixteenth- and seventeenth-century Italian mathematics is discussed in a series of papers:

Bortolotti, E. A series of papers written between 1922 and 1928, e.g., in *Periodico di matematica*, Vol. 5 (1925), pp. 147–84; *ibid.*, Vol. 6 (1926), pp. 217–30; *ibid.*, Vol. 8 (1928), pp. 19–59; *Scientia* (1923), pp. 385–94.
——. *I contributi del Tartaglia, del Cardano, del Ferrari, e della scuola matematica bolognese alla teoria algebrica delle equazioni cubiche*. Imola, 1926.

Cardano's autobiography has been translated:

Cardano, J. *The Book of My Life*. London, 1931. (Translation by J. Stoner.)

See also:

Ore, O. *Cardano, the Gambling Scholar*. Princeton, 1953.

Much information about sixteenth- and seventeenth-century mathematicians and their works is given in the papers of H. Bosmans, S. J., most of which can be found in the *Annales de la Société Scientifique Bruxelles*, 1905–27. (A complete list is given by A. Rome in *Isis*, Vol. 12 [1929], p. 88–112.)

Modern text editions or translations of authors discussed in this chapter are:

Grant, E., ed. *De proportionibus proportionum* (by Nicole Oresme). Madison, Wis., 1966.
Boncampagni, B., ed. *Scritti di Leonardo Pisano*. 2 vols., Rome, 1857–62.
Hughes, B., trans. *Regiomontanus on Triangles*. Madison, Wis., etc., 1967. See B. Rosenfeld, *Scripta math.*, Vol. 28 (1970), pp. 364–65.
Gould, S. H., trans. *The Book on Games of Chance* (by Girolamo Cardano). New York, 1961.
Stevin, S. *The Principal Works*. 5 vols., Amsterdam, 1955–66. Vols. IIA, B (1958) contain the Mathematics.
Masotti, A., ed. *Quaesiti* (by Tartaglia) and *Cartelli* (by Tartaglia and Ferrari). Brescia, 1959, 1974.

Viète, F. *Opera mathematica*. Leiden, 1646. Reprinted with preface by J. E. Hofmann, Hildesheim and New York, 1970.

Macdonald, W. R., trans. *The Construction of the Wonderful Canon of Logarithms* (by Napier). Edinburgh, 1889.

Busard, H. L. L., ed. *Quaestiones Super Geometriam Euclidis* (by Nicole Oresme). Leiden, 1961.

Hofmann, J. and J. E., trans. and ed. *Mathematische Schriften* (by Nikolaus von Cues). Hamburg, 1952.

Crosby, H. L., ed. *Tractatus de Proportionibus* (by Thomas of Bradwardine). Madison, Wis., 1955.

Arrighi, G., ed. *Trattato d'aritmetica* (by Piero Dell'Abbaco). Pisa, 1964.

Bubnow, N., ed. *Gerberti postea Silvestri II papae Opera Mathematica*. Berlin, 1899. New ed. with added material, Hildesheim, 1913.

Witmer, T. R., trans. and ed. *The Great Art or the Rules of Algebra by Girolamo Cardano*. Cambridge, Mass., and London, 1968. (This is the "Ars Magna" in English with modern notation.)

Furthermore, see:

Carslaw, H. S. "The Discovery of Logarithms by Napier." *Math. Gazette*, Vol. 8 (1915–16), pp. 76–84, 115–19.

Blaschke, W., and Schoppe, G. *Regiomontanus, Commensurator*. Berlin, 1956. *Napier Tercentenary Memorial Volume*, C. G. Knott, ed. London and New York, 1915.

Zinner, E. *Leben und Wirken des Johannes Müller von Königsberg genannt Regiomontanus*. Munich, 1938.

Bond, J. D. "The Development of Trigonometric Methods Down to the Close of the Fifteenth Century." *Isis*, Vol. 4 (1921–22), pp. 295–323.

Yeldham, F. A. *The Story of Reckoning in the Middle Ages*. London, 1926.

Dijksterhuis, E. J. *Simon Stevin*. The Hague, 1943. (In English: The Hague, 1970.)

Thorndike, L. *The Sphere of Sacrobosco*. Chicago, 1949.

Geyer, B. "Die mathematischen Schriften des Albertus Magnus." *Angelicus*, 35 (1958), pp. 159–75.

Bodewig, E. "Die Stellung des heiligen Thomas von Aquino zur Mathematik." *Arch. f. Geschichte der Philosophie*, Vol. 11 (1931), pp. 1–34.

Hofmann, J. E. "Über Viète's Beiträge zur Geometrie der Einschiebungen." *Mathphysik, Semesterberichte* 8 (1962), pp. 191–214.

Taylor, E. G. R. *The Mathematical Practitioners of Tudor and Stuart England*. Cambridge, 1954.

Voelling, E. "Jost Bürgi und die Logarithmen." In *Elemente der Mathematik*, Suppl. 5, Basel, 1948.

Treutlein, P. "Das Rechnen im 16. Jahrhundert." *Abh. zur Geschichte der Mathematik*, 1 (1877), pp. 1–100.

——. "Die deutsche Coss." *Ibid.*, 2 (1879), pp. 1–124.

Clagett, M. *The Science of Mechanics in the Middle Ages*. Madison, Wis., and London, 1959.

——. *Archimedes in the Middle Ages*, Vol. I. Madison, Wis., 1964.

Sarton, G. *Six Wings: Men of Science of the Renaissance*. Bloomington, Ind., 1957.

Averdunk, H., and Müller, Reinhard J. *Gerhard Mercator*. In *Petermanns Mitteilungen*, Ergänzungsheft 182. Gotha, 1914, 100 pp.

Davis, N. Z. "Sixteenth Century French Arithmetics and the Business Life." *J. Hist. of Ideas*, Vol. 21 (1960), pp. 18–48.

Bockstaele, P. "Adriaan van Roomen," *National Biogr. Woordenbock* 2. Brussels, 1966. Pp. 752–65.

Smith, D. E. *Rara Arithmetica*. Boston and London, 1908. Fourth ed., New York, 1970.

Rose, P. L. *The Italian Renaissance of Mathematics*. Geneva, 1975.

Matvierskaya, G. P. *Development of Number Theory in Europe till the 17th Century* (Russian). Tashkent, 1971.

CHAPTER VI

The Seventeenth Century

1

The rapid development of mathematics during the Renaissance was due not only to the *Rechenhaftigkeit* of the commercial classes but also to the productive use and further perfection of machines. Machines were known to the Orient and to classical antiquity; they had inspired the genius of Archimedes. However, the existence of slavery and the absence of an economically progressive urban life frustrated the use of machines in these older forms of society. This is indicated by the works of Heron, where machines are described, but only for the purpose of amusement or deception.

In the later Middle Ages machines came into use in small manufactures, in public works, and in mining. These were enterprises undertaken by city merchants or by princes in search of ready money and often conducted in opposition to the city guilds. Warfare and navigation also stimulated the perfection of tools and their further replacement by machines.

A well-established silk industry existed in Lucca and in Venice as early as the fourteenth century. It was based on division of labor and on the use of water power. In the fifteenth century, mining in Central Europe developed into a completely capitalistic industry based technically on the use of pumps and hoisting machines which allowed the boring of deeper and deeper layers. The invention of firearms and of printing, the construction of windmills and canals, the building of ships to sail the ocean, required engineering skill and made people technically conscious. The perfection of clocks, useful for astronomy and navigation and often installed in public places, brought admirable pieces of mechanism before the public eye; the regularity of their motion and the possibility they offered of indicating time exactly made a deep impression upon the philosophical mind. During the Renaissance, and even centuries later, the clock was taken as a model of the universe. This was an important factor in the development of the mechanical conception of the world. It also represented a psychological change, expressed in the words "time is money."

Machines led to theoretical mechanics and to the scientific study of motion and of change in general. Antiquity had already produced texts on statics, and the new study of theoretical machines naturally based itself upon the statics of the classical

93

authors. Books on machines appeared long before the invention of printing, first empirical descriptions (Kyeser, early fifteenth century), later more theoretical ones, such as Leon Battista Alberti's book on architecture (*c.* 1450) and the writings of Leonardo da Vinci (*c.* 1500). Leonardo's manuscripts contain the beginnings of a definite mechanistic theory of nature. Tartaglia, in his *Nuova scienzia* (1537), discussed the construction of clocks and the orbit of projectiles— but had not yet found the parabolic orbit, first discovered by Galileo. The publication of Latin editions of Heron and Archimedes stimulated this kind of research, especially F. Commandino's edition of Archimedes, which appeared in 1558 and brought the ancient method of integration within the reach of the mathematicians. Commandino himself applied these methods to the computation of centers of gravity (1565), though with less rigor than his master.

This computation of centers of gravity remained a favorite topic of Archimedean scholars, who used their study of statics to obtain a working knowledge of the rudiments of what we now recognize as the calculus. Outstanding among such students of Archimedes were Simon Stevin, who wrote on centers of gravity and on hydraulics, both in 1586; Luca Valerio, who wrote on centers of gravity in 1604 and on the quadrature of the parabola in 1606; and Paul Guldin, in whose *Centrobaryca* (1641) we find the so-called theorem of Guldin on centroids, already explained by Pappus. In the wake of the early pioneers came the great works of Kepler, Cavalieri, Torricelli, and others, in which were evolved methods which eventually led to the invention of the calculus.

2

Typical of these authors was their willingness to abandon Archimedean rigor for considerations often based on nonrigorous, sometimes "atomic," assumptions (probably without knowing that Archimedes, in his letter to Eratosthenes, had also used such methods for their heuristic value). This was partly due to impatience with scholasticism among some, though not all, of these authors, since several of the pioneers were Catholic priests trained in scholasticism. The main reason was the desire for results, which the Greek method was unable to provide quickly.

The revolution in astronomy, connected with the names of Copernicus, Tycho Brahe, and Kepler, opened entirely new visions of man's place in the universe and man's power to explain the phenomena of astronomy in a rationalistic way. The possibility of a celestial mechanics to supplement terrestrial mechanics increased the boldness of the men of science. In the works of Johann Kepler the stimulating influence of the new astronomy on problems involving large computations as well as infinitesimal considerations is particularly evident. Kepler even ventured into volume computation for its own sake, and, in his *Nova stereometria doliorum vinariorum* ("New solid geometry of wine barrels"; 1615) evaluated the volumes of solids obtained by rotating segments of conic sections about an axis in their plane. He broke with Archimedean rigor; his circle area was composed of an infinity of triangles with common vertex at the center; his sphere consisted of an infinity of pointed pyramids. The proofs of Archimedes, Kepler said, were absolutely rigorous, *absolutae et omnibus numeris perfectae*,[1] but he left them to the people who wished to indulge in exact demonstrations. Each successive author was free to find his own kind of rigor, or lack of rigor, for himself.

To Galileo Galilei we owe the new kinematics of freely falling bodies, the beginning of the theory of elasticity, and a spirited defense of the Copernican system. Above all we owe to Galileo, more than to any other man of his period, the spirit of modern science based on the harmony of experiment and theory, with stress on the intensive use of mathematics (although there is less experiment in Galileo than is sometimes believed). In the *Discorsi e dimostrazioni matematiche intorno a due nuove scienze* (1638, the two new sciences being "mechanics and local motions"), Galileo was led to the mathematical study of motion, to the relation between distance, velocity, and acceleration. He never gave a systematic explanation of his ideas on the calculus, leaving this to his pupils Torricelli and Cavalieri. Indeed, Galileo's ideas on these questions of pure mathematics were quite original, as appears from his remark that "neither is the number of squares less than the totality of all numbers, nor the latter greater than the former." This defense of the actually infinite (given by Salviati in the *Discorsi*) was consciously directed against the Aristotelian and scholastic position (represented by Simplicio). The *Discorsi* also contain the parabolic orbit of the projectile, with tables for height and range as functions of the angle of elevation and given initial velocity. Salviati also remarks that the catenary looks like a parabola, but does not give the precise description of the curve.

Galileo wrote in Italian, as Stevin did in Dutch, Bacon in English, and Descartes in French, for these men wanted to reach that larger public which by this time was ready to learn about the new sciences.

The time had now arrived for a first systematic exposition of the results reached so far in what we now call the calculus. This exposition appeared in the *Geometria indivisibilibus continuorum* (1635) of Bonaventura Cavalieri, professor at the University of Bologna. Here Cavalieri established a simple form of the calculus, basing it on the scholastic conception of the *indivisible*,[2] the point generating the line, the line generating the plane by motion. He then added line segments to obtain an area, plane segments to obtain a volume; but when Torricelli once showed him that in this way one can prove that any triangle is divided by an altitude into two parts of equal area, he replaced "lines" by "threads," hence lines of a certain small breadth, and thus came to an "atomic" theory. His ideas on lines building up an area led him to the correct "principle of Cavalieri," which concludes that two solids of equal altitudes have the same volume, if plane cross-sections at equal height have the same area. It allowed him to perform the equivalent of the integration of polynomials.

---

[1] "Absolute and in all respects perfect."

[2] F. Cajori, "Indivisibles and 'ghosts of departed quantities' in the history of mathematics," *Scientia*, Vol. 37 (1925), pp. 301–06; E. Hoppe, "Zur Geschichte der Infinitesimalrechnung bis Leibniz und Newton," *Jahresb. Deutsch. Math. Verein*, Vol. 37 (1928), pp. 148–87. On certain statements in Hoppe see C. B. Boyer, *The History of the Calculus* (New York, 1949), pp. 192, 206, 209. (Dover reprint, 1959.)

3

The evolution of the calculus, part of the whole renewal of mathematics, occurred in an intellectual climate in which thinkers were gradually outgrowing the ancient Aristotelian outlook on nature. Aristotle, emphasizing qualities and teleology, was felt to be utterly inadequate in a world where measurement, computation, engineering, and quantity in general, with their causal relationships, were becoming more and more important. A new method of approach to nature and man was felt necessary in leading circles, and with it a new mathematics, what was to become the classical example of quantitative and logical thinking.

This gradual evolution was considerably stimulated by the publication of Descartes's *Géométrie* (1637), which brought the whole field of classical geometry within the scope of the algebraists. The book was originally published as an appendix to the *Discours de la Méthode*, the discourse on reason, in which the author explained his rationalistic approach to the study of nature. René Descartes was a Frenchman from Touraine who lived the life of a gentleman. He served for a while in the army of Maurice of Orange, resided for many years in the Netherlands, and died in Stockholm, to which city he had been invited by the Queen of Sweden. In accordance with many other great thinkers of the seventeenth century, Descartes searched for this general method of thinking in order to be able to facilitate inventions and "to find the truth in the sciences." Since the only known natural sciences with some degree of systematic coherence were astronomy and mechanics, and the key to the understanding of mechanics and astronomy was mathematics, mathematics became the most important means for the understanding of the universe. Moreover, mathematics with its convincing statements was itself the brilliant example that truth could be found in science. The mechanistic philosophy of this period thus came to a conclusion that was similar to that of the Platonists, but for a different reason. Platonists, believing in the harmony of the universe, and Cartesians, believing in a general method based on reason, both found in mathematics the queen of the sciences.

Descartes published his *Géométrie* as an application of his general method of rationalistic unification, in this case the unification of algebra and geometry. The merits of the book, according to the commonly accepted point of view, consist mainly in the creation of so-called analytic geometry. It is true that this branch of mathematics eventually evolved under the influence of Descartes's book, but the *Géométrie* itself can hardly be considered a first textbook on this subject. There are no explicit "Cartesian" axes, and no equations of the straight line or of conic sections are derived, though particular equations of the second degree are interpreted as denoting a conic section. Moreover, a large part of the book consists of a theory of algebraic equations, containing the "rule of Descartes" to determine the number of positive and negative roots (he called them true and false roots).

We must keep in mind that Apollonius already had a characterization of conic sections by means of what we now—with Leibniz—might call coordinates, even though no numerical values were attached to it. However, latitude and longitude in Ptolemy's *Geographia* were numerical coordinates. Pappus, in his *Collection*,

JOHANN KEPLER (1571–1630)

GALILEO GALILEI (1564–1642)
(From a painting in the Pitti
Palace, Florence, school of J.
Sutterman(s), painter of the
Medici.)

RENÉ DESCARTES (1596–1650)
(From a painting by Frans Hals.)

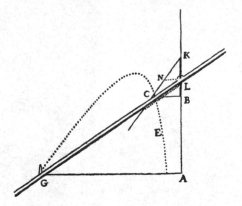

Aprés cela prenant vn point a diſcretion dans la courbe,
comme C, ſur lequel ie ſuppoſe que l'inſtrument qui ſert
a la deſcrire eſt appliqué, ie tire de ce point C · la ligne
C B parallele a G A, & pourceque C B & B A ſont deux
quantités indeterminées & inconnuës , ie les nomme
l'vne *y* & l'autre *x*. mais affin de trouuer le rapport de
l'vne à l'autre ; ie conſidere auſſy les quantités connuës
qui determinent la deſcription de céte ligne courbe,
comme G A que ie nomme *a*, K L que ie nomme *b*, &
N L parallele a G A que ie nomme *c*. puis ie dis, comme
N L eſt à L K, ou *c* à *b*, ainſi C B, ou *y*, eſt à B K, qui eſt
par conſequent $\frac{b}{c}\,y$: & B L eſt $\frac{b}{c}\,y - b$, & A L eſt *x* +

$\frac{b}{c}\,y - b$. de plus comme C B eſt à L B, ou *y* à $\frac{b}{c}y - b$, ainſi
a, ou G A, eſt á L A, ou *x* $+ \frac{b}{c}y - b$. de façon que mul-

S ſ   tipliant

A FACSIMILE PAGE OF DESCARTES'S "LA GÉOMÉTRIE."[3]

[3]A ruler AK of indefinite length is fixed; a ruler GL, also of indefinite length, can turn around a
pivot G; and a triangular figure KCB has a side KC, again of indefinite length. The ruler GL is
hinged at L. The figure KCB is made to slide along the ruler GL, KB moving on the fixed "axis"
AB. The locus of C when GL turns around G is found to be

$$y^2 = cy - \frac{cx}{B}y + ay - ac.$$

(Descartes writes *yy* and has a special sign for =.) Hence a hyperbola.

had a "Treasury of Analysis" (*Analuomenos*) in which we have only to modernize the notation to obtain an application of algebra to geometry. Even a glimpse of a graphical representation occurs occasionally before Descartes, for example, in the work of Oresme. Descartes's merits lie above all in his consistent application of the well-developed algebra of the sixteenth century to the geometrical analysis of the ancients, and by this, in an enormous widening of its applicability. A second merit is Descartes's final rejection of the homogeneity restrictions of his predecessors, which even vitiated Viète's *logistica speciosa*, so that x^2, x^3, xy were now considered as line segments. An algebraic equation became a relation between numbers, a new advance in mathematical abstraction. This then was utilized for the further development of algebra and for the general treatment of algebraic curves. The West, in catching up with the Oriental arithmetical-algebraic tradition, was fast surpassing it.

Much in Descartes's notation is already modern; we find in his book expressions such as

$$\tfrac{1}{2}a + \sqrt{\tfrac{1}{4}aa + bb},$$

which differs from our own notation only in Descartes's still writing aa for a^2 (which is even found in Gauss), though he has a^3 for aaa, a^4 for $aaaa$, etc. It is not hard to find one's way in his book, but we must not look for our modern analytic geometry.

A little closer to such analytic geometry came Pierre Fermat, a lawyer at Toulouse, who wrote a short paper on geometry probably before the publication of Descartes's book, but which was only published in 1679. In this *Isagoge* we find the equations

$$y = mx, \ xy = k^2, \ x^2 + y^2 = a^2, \ x^2 \pm a^2y^2 = b^2$$

assigned to lines and conics, with respect to a system of (usually perpendicular) axes. However, since it was written in Viète's notation, the paper looks more archaic than Descartes's *Géométrie*. At the time Fermat's *Isagoge* was printed, there were already other publications in which algebra was applied to Apollonius' results, notably the *Tractatus de sectionibus conicis* (1655), by John Wallis, and a part of the *Elementa curvarum linearum* (1659), written by Johan De Witt, grand pensionary of Holland. Both these works were written under the direct influence of Descartes. But progress was very slow; even L'Hospital's *Traité analytique des sections coniques* (1707) has not much more than a transcription of Apollonius into algebraic language. All authors hesitated to accept negative values for the coordinates. The first to work boldly with algebraic equations was Newton in his study of cubic curves (1703); the first analytic geometry of conic sections that is fully emancipated from Apollonius appeared only with Euler's *Introductio* (1748).

4

The appearance of Cavalieri's book stimulated a considerable number of mathematicians in different countries to study problems involving infinitesimals. The fundamental problems began to be approached in a more abstract form and in this way gained in generality. The tangent problem, consisting in the search for methods to find a tangent to a given curve at a given point, took a more and more prominent place beside the ancient problems involving volumes and centers of gravity. In this search there were two marked trends, a geometrical and an algebraic one. The followers of Cavalieri, notably Torricelli and Isaac Barrow, Newton's teacher, loved the Greek method of geometrical reasoning without caring too much about its rigor. Christiaan Huygens also showed a definite partiality for Greek geometry. There were others, notably Fermat, Descartes, and John Wallis, who showed the opposite trend and brought the new algebra to bear upon the subject. Practically all authors in this period from 1630 to 1660 confined themselves to questions dealing with algebraic curves, especially those with equation $a^m y^n = b^n x^m$. And they found, each in his own way, formulas equivalent to $\int_0^a x^m \, dx = a^{m+1}/(m + 1)$, first for positive integer m, later for m negative integer and fractional.[4] Occasionally a nonalgebraic curve appeared, such as the cycloid (roulette) investigated by Descartes and Blaise Pascal; Pascal's *Traité général de la roulette* (1658), a part of a booklet published under the name of A. Dettonville, had great influence on young Leibniz.[5]

In this period several characteristic features of the calculus began to appear. Fermat discovered in 1638 a method to find maxima and minima by slightly changing the variable in a simple algebraic equation and then letting the change disappear; it was generalized in 1658 to more general algebraic curves by Johannes Hudde, later a mayor of Amsterdam. There were determinations of tangents, volumes, and centroids, but the relation between integration and differentiation as inverse problems was not really grasped until Barrow explained it in 1670, although in a difficult geometrical form. Pascal occasionally used expansions in terms of small quantities in which he dropped the terms of lower dimensions—anticipating the debatable assumption of Newton that $(x + dx)(y + dy) - xy = xdy + ydx$. Pascal defended his procedure by appealing to intuition (*esprit de finesse*) rather than to logic (*esprit de géométrie*), here anticipating Bishop Berkeley's criticism of Newton.[6]

Scholastic thought entered into this search for new methods not only through Cavalieri but also through the work of the Belgian Jesuit Grégoire de Saint-Vincent and his pupils and associates, Paul Guldin and André Tacquet. These men were inspired by both the spirit of their age and the medieval scholastic writings on the nature of the continuum and the latitude of forms. In their writings the term "exhaustion" for Archimedes' method appears for the first time. Tacquet's book, *On Cylinders and Rings* (1651; in Latin), influenced Pascal.

This fervid activity of mathematicians in a period when no scientific periodicals existed led to discussion circles and to constant correspondence. Some figures gained merit by serving as a center of scientific interchange. The best-known of

---

[4] The case m = -1 offered special difficulties, which were only overcome when the relationship between logarithms and exponentials was fully understood.

[5] H. Bosmans, "Sur l'oeuvre mathématique de Blaise Pascal," *Revue des Questions Scient.*, Vol. 4, Ser. 5 (1929), pp. 136–60, 424–50; J. Guitton, *Pascal et Leibniz* (Paris, 1951).

[6] B. Pascal, *Oeuvres* (Paris, 1908–14), Vol. XII, p. 9; Vol. XIII, pp. 141–55.

these men is the Minorite Father Marin Mersenne, whose name as a mathematician is preserved in "Mersenne's numbers." Descartes, Fermat, Desargues, Pascal, and many other scientists corresponded with him.[7] Academies crystallized out of the discussion groups of learned men. They arose, in a way, as opposition to the universities, which had maintained the spirit of the scholastic period—with some exceptions such as Leiden University—and which fostered the medieval attitude of presenting knowledge in fixed forms. The new academies, on the contrary, expressed the new spirit of investigation. As one writer observes, they typified

> . . . this age drunk with the fulness of new knowledge, busy with the uprooting of superannuated superstitions, breaking loose from traditions of the past, embracing most extravagant hopes for the future. Here the individual scientist learned to be contented and proud to have added an infinitesimal part to the sum of knowledge; here, in short, the modern scientist was evolved.[8]

The first academies of the Renaissance were founded in Italy; they were of a literary-philosophical nature. A scientific academy, mentioned at Naples some time before 1580, anticipated the Accademia dei Lincei (lynxes have sharp sight), founded in 1603 and still extant after several interruptions. The Royal Society of London dates from 1662, the French Académie des Sciences from 1666. Wallis was a charter member of the Royal Society, Huygens of the French Academy.

5

Next to that of Cavalieri among the most important books written in this period of anticipation is Wallis' *Arithmetica infinitorum* (1655). The author, from 1643 until his death in 1703, was the Savilian professor of geometry at Oxford. The very title of his book shows that Wallis intended to go beyond Cavalieri with his *Geometria indivisibilibus*; it was the new *arithmetica* (algebra) which Wallis wanted to apply, not the ancient geometry. In this process Wallis extended algebra into a veritable analysis—the first mathematician to do so. His methods of dealing with infinite processes were often crude, but he obtained new results; he introduced infinite series and infinite products and used with great boldness imaginaries and negative and fractional exponents. He wrote ∞ for $\frac{1}{0}$ (and claimed that $-1 > \infty$). Typical of his results is the expansion

$$\frac{\pi}{2} = \frac{2 \cdot 2 \cdot 4 \cdot 4 \cdot 6 \cdot 6 \cdot 8 \cdot 8 \cdots}{1 \cdot 3 \cdot 3 \cdot 5 \cdot 5 \cdot 7 \cdot 7 \cdot 9 \cdots},$$

as well as expressions equivalent to Beta integrals. For what we now express by $4/\pi$ he wrote a little square.

Wallis was only one of a whole line of brilliant men of this period who enriched

---

[7]*Informer Mersenne d'une découverte, c'était la publier par l'Europe entière*, writes H. Bosmans (*op. cit.*, p. 43: "To inform Mersenne of a discovery, meant to publish it throughout the whole of Europe"). The convent at which Mersenne taught was at the present Place des Vosges in Paris, and here his visitors came. "Le bon père Mersenne" died in 1648.

[8]M. Ornstein, *The Role of Scientific Societies in the Seventeenth Century* (Chicago, 1913), p. 262.

mathematics with discovery after discovery. The driving force behind this flowering of creative science, unequaled since the great days of Greece, was only in part the ease with which the new techniques could be handled. As already mentioned, many great thinkers were in search of more: of a "general method"—sometimes conceived in a restricted sense as a method of mathematics, sometimes more general as a method of understanding nature and of creating new inventions. This is the reason why in this period all outstanding philosophers were mathematicians, and all outstanding mathematicians were philosophers. The search for new inventions sometimes led directly to mathematical discoveries. A famous example is the *Horologium oscillatorium* (1673) of Christiaan Huygens, where the search for better timepieces (in order to solve the age-old problem of finding longitude at sea) led not only to pendulum clocks, but also to the study of evolutes and involutes of a plane curve. Huygens, who was a Hollander of good family, resided for many years in Paris. He was a leading force in the newly founded Académie des Sciences.[9] Eminent as a physicist as well as an astronomer, he established the wave theory of light and explained that Saturn had a ring. His book on pendulum clocks was of influence on Newton's theory of gravitation; it represents, with Wallis' *Arithmetica*, the most advanced form of the calculus in the period before Newton and Leibniz. The letters and books of Wallis, Huygens, and their colleagues abound in new discoveries, in rectifications, envelopes, and quadratures. Huygens studied the tractrix, the logarithmic curve, and the catenary, and established the cycloid as a tautochronous curve. Despite this wealth of results, many of which were found after Leibniz had published his calculus, Huygens belongs definitely to the period of anticipation. He confessed to Leibniz that he never was able to familiarize himself with Leibniz's method. Wallis, in the same way, never found himself at home in Newton's notation. Huygens was one of the great seventeenth-century mathematicians who took rigor seriously; his methods were always in the Archimedean tradition.

6

The activity of the mathematicians of this period stretched into many fields, new and old. They enriched classical topics with original results, cast new light upon ancient fields, and even created entirely new subjects of mathematical research. An example of the first case was Fermat's study of Diophantus; an example of the second case was Desargues's new interpretation of geometry. The mathematical theory of probability was an entirely new creation.

Diophantus became available to a Latin-reading public in 1621.[10] In Fermat's copy of this translation are found his famous marginal notes, which his son published in 1670. Among them we find Fermat's "great theorem" that $x^n + y^n =$

---

[9]The Académie Française was instituted by Richelieu in 1635; its principal function was the composition of a dictionary of the French language. Its forty members (to which scientists were also elected) are the "immortals." This academy was followed by others, among which was the Académie des Sciences (1666). These academies were reorganized at the time of the French Revolution in the Institut National, to which the Académie des Sciences belongs.

[10]First readily available Latin translations: Euclid, 1482; Ptolemy, 1515; Archimedes, 1558; Proclus, 1560, Apollonius I–IV, 1566, V–VII, 1661; Pappus, 1589; Diophantus, 1621.

z^n is impossible for positive integer values of x, y, z, n if $n > 2$, which led Kummer in 1847 to his theory of ideal numbers. A proof valid for all n has not yet been given, though the theorem is certainly correct for a large number of values.[11]

Fermat wrote in the margin beside Diophantus II, 8: "To divide a square number into two other square numbers," the following words:

> To divide a cube into two other cubes, a fourth power, or in general any power whatever into two powers of the same denomination above the second is impossible, and I have assuredly found an admirable proof of this, but the margin is too narrow to contain it.

If Fermat had such admirable proof, then three centuries of intense research have failed to produce it again. It is safer to assume that even the great Fermat slept sometimes.

Another marginal note of Fermat states that a prime of form $4n + 1$ can be expressed once, and only once, as the sum of two squares, which theorem was later demonstrated by Euler. The other "theorem of Fermat," which states that $a^{p-1} - 1$ is divisible by p when p is prime and a is prime to p, appears in a letter of 1640; this theorem can be demonstrated by elementary means. Fermat was also the first to assert that the equation $x^2 - Ay^2 = 1$ (A a nonsquare integer) has an unlimited number of integer solutions (1657).

Fermat and Pascal were the founders of the mathematical theory of probabilities. The gradual emergence of the interest in problems relating to probabilities is primarily due to the development of insurance, but the specific questions which stimulated great mathematicians to think about this matter came from requests of noblemen gambling in dice or cards. In the words of Poisson: *Un problème relatif aux jeux de hasard, proposé à un austère janséniste par un homme du monde, a été l'origine du calcul des probabilités.*[12] This "man of the world" was the Chevalier de Méré (a gentleman of considerable erudition), who approached Pascal with a question concerning the so-called *problème des points*. Pascal began a correspondence with Fermat on this problem and on related questions, and both men established some of the foundations of the theory of probability (1654). When Huygens came to Paris he heard of this correspondence and tried to find his own answers; the result was the *De ratiociniis in ludo aleae* (1657), the first treatise on probability based on the concept of expectation.[13] The next steps were taken by De Witt and Halley, who constructed tables of annuities (1671, 1693).

Blaise Pascal was the son of Étienne Pascal, a correspondent of Mersenne; the "limaçon of Pascal" is named after Étienne. Blaise developed rapidly under his father's tutelage and at the age of sixteen discovered "Pascal's theorem" concerning a hexagon inscribed in a circle. It was published in 1641 on a single sheet of

---

[11]See P. Bachman, *Der Fermatsche Satz*, Berlin, 1919; H. S. Vandiver, *Amer. Math. Monthly*, Vol. 53 (1946), pp. 555–78; O. Ore, *Number Theory and Its History*, New York, 1948; L. J. Mordell, *Three Lectures on Fermat's Last Theorem* (Cambridge, 1921), reprinted Berlin, 1972; H. M. Edwards, *Fermat's Last Theorem* (New York, 1974).

[12]"A problem concerning games of chance, proposed by a man of the world to an austere Jansenist, was the origin of the calculus of probabilities" (S. D. Poisson, *Recherches sur la probabilité des jugements*, Paris, 1837, p. 1).

[13]H. Freudenthal, "Huygens' Foundation of Probability," *HM*, Vol. 7 (1980), pp. 113–17.

CHRISTIAAN HUYGENS (1629–1695)
(From an engraving by Gerald Edelinck, a Flemish etcher,
after a drawing by the French etcher Pierre Drevet.)

BLAISE PASCAL (1623–1662)

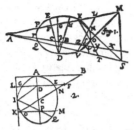
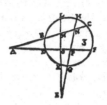

ESSAY POVR LES CONIQVES.   Par B. P.

DEFINITION PREMIERE.

Vand plusieurs lignes droites concourent à mesme point, ou sont toutes paralleles entr'elles, toutes ces lignes sont dites de mesme ordre ou de mesme ordonnance, & la multitude de ces lignes, est dite ordre de lignes, ou ordonnance de lignes.

DEFINITION II.

Par le mot de section de Cone nous entendons la circonference du Cercle, l'Elipse, l'Hyperbole, la Parabole & la ligne rectiligne, d'autant qu'vn Cone coupe parallelement à sa base, ou par son sommet ou des trois autres sens qui engendrent l'Elipse, l'Hyperbole & la parabole engendre dans la superficie Conique, ou la circonference d'vn Cercle ou vn Angle, ou l'Elipse, ou l'Hyperbole, ou la parabole.

DEFINITION III.

Par le mot de droite mis seul, nous entendons ligne droite.

LEMM. I.

Figure. I. Si dans le plan, M, S, Q, du point M partent les deux droites M K, M V, & du point S, partent les deux droites S K, S V, & que K, soit le concours des droites M K, S K, & V, le concours des droites, M V, S V, & A, le concours des droites M A, S A, & μ, le concours des droites M V, S K, & que par deux des quatre points, A K à V. qui ne soient point en mesme droite auec les points, M, S, comme par les points, K, V, passe la circonference d'vn cercle coupante les droites M V, M P, S V, S K, és points, O, P, Q, N, ie dis que les droites, M S, N O, P Q, sont de mesme ordre.

LEMM. II.

Fig. I. Si par la mesme droite passent plusieurs plans, qui soient coupez par vne autre plan, toutes les lignes des sections de ces plans sont de mesme ordre auec la droite par laquelle passent lesdits plans.

Fig. I. Ces deux Lemmes posez & quelques faciles consequences d'iceux nous demonstrerons que les mesmes choses estant posées qu'au premier Lemme, si par les points, K, V, passe vne quelconque section de Cone qui coupe les droites M K, M V, S K, S V, és points, P, O, N, Q, les droites M S, N O, P Q, seront de mesme ordre, cela vaut vn troisiéme Lemme.

En suite de ces trois Lemmes & de quelques consequences d'iceux nous donnerons des Elemens Coniques complets, à sçauoir toutes les proprietez des diametres & costez droits, destangentes &c. la restitution du Cone presque sur toutes les données, la description des sections de Cone par points, &c.

Fig. I. Quoy faisans, nous enonçons les proprietez que nous en touchons d'vne maniere plus vniuerselle qu'à l'ordinaire. Par exemple celle cy, si dans le plan M S Q, dans la section de Cone, P K V, sont menées les droites A K, A V, atteignantes la section aux points P K, Q V, & que de deux des quatre points qui ne sont point en mesme droite auec le point A, comme par les points K, V, & par deux points N, O, pris dans le bord de la section sont menées quatre droites K N, K O, V N, V O, coupantes les droites A V, A P, aux points L, M, T, S, ie dis que la raison composée des raisons de la droite P M, à la droite M A, & de la droite A S, à la droite S Q, est la mesme que la composée des raisons de la droite P L, à la droite L A, & de la droite A T, à la droite T Q.

Fig. I. Nous demonstrerons aussi que s'il y a trois droites D E, D G, D H, que les droites A P, A R, coupent aux points F, G, H, C, γ, B, & que dans la droite D C, soit determiné le point E, la raison composée des raisons du rectangle E F, en F G, au rectangle de E C, en C γ, & de la droite A γ, à la droite A G, est la mesme que la composée des raisons du rectangle E F, en F H, au rectangle E C, en C B, & de la droite A B, à la droite A H. Et est aussi la mesme que la raison du rectangle des droites F E, F D, & rectangle des droites, C E, C D, partant si par les points E, D, passe vne section de Cone qui coupe les droites A H, A B, és points P, K, R, ψ, la raison composée des raisons du rectangle des droites E F, F C, en rectangle des droites P M, à la droite M A, & de la droite A S, à la droite S Q, est la mesme que la composée des raisons du rectangle des droites F K, F P, au rectangle de droites C R, C ψ, & de rectangle des droites A R, A ψ, au rectangle des droites A K, A P.

Fig. III. Nous demonstrerons aussi que si quatre droites A C, A F, E H, E L, s'entrecoupent és points N, P, M, O, & qu'vne section de Cone coupe lesdites droites és points C, B, E, D, H, G, L, K, la raison composée des raisons du rectangle de M C, en M B, au rectangle des droites P F, P D, & de rectangle des droites A D, A F, au rectangle des droites A B, A C, est la mesme que la raison composée des raisons du rectangle des droites M L, M K, au rectangle des droites P H, P G, & de rectangle des droites E H, E G, au rectangle des droites E K, E L.

Fig. I. Nous demonstreron aussi cette proprieté, dont le premier inuenteur est M.r Desargues Lyonnois, vn des grands esprits de ce temps, & des plus versez aux Mathematiques, & entr'autres aux Coniques, dont les escripts sur cette matiere, quoy qu'en petit nombre, en ont donné vn ample tesmoignage à ceux qui en auront voulu receuoir l'intelligence: & veux bien aduouer que ie doibs le peu que i'ay trouué sur cette matiere à ses escrits, & que i'ay tasché d'imiter autant qu'il m'a esté possible sa methode sur ce suiet, qu'il a traité sans se seruir du triangle par l'axe. Et traittant generalement de toutes les sections de Cone, la proprieté merueilleuse dont est question est telle: si dans le plan M S Q, y a vne section de Cone P Q V, dans le bord de laquelle ayant pris les quatre points K, N, O, V, sont menées les droites K N, K O, V N, V O, de sorte que par vn mesme des quatre points ne passent que deux droites, & qu'vne autre droite coupe tant l'abord de la section aux points R, ψ, que les droites K N, K O, V N, V O, és points x y Z ψ, ie dis que comme le rectangle des droites Z r, Z ψ, est au rectangle des droites y r, y ψ, ainsi le rectangle des droites x r, x ψ, est au rectangle des droites x r, x ψ.

Fig. II. Nous demonstrerons aussi que si dans le plan de l'hyperbole, ou du cercle A G E, dont le centre est C, on mene la droite A B, touchant au point A, la section, & qu'ayant mené le diametre C A, on prene la droite A B, dont le quarré soit égal au quart du rectangle de la figure, & qu'on mene C B, alors quelque droite qu'on mene D E, parallele à la droite A B, coupante la section en E, & les droites A C, C B, és points D F, si la section A G E, est vne elipse ou vn cercle, la somme des quarrez des droites D E, D F, sera egale aux quarré de la droite A B, & dans l'hyperbole la difference des mesmes quarrez des droites D E, D F, sera egale au quarré de la droite A B.

Nous deduirons aussi quelques problemes, par exemple d'vn poinct donné mener vne droite touchant vne section de Cone donnée.

Trouuer deux diametres coniuguez en angle donné.

Trouuer deux diametres en angle donné & en raison donnée.

Nous tirerons plusieurs autres Problemes & Theoremes & plusieurs consequences des precedents, mais la defiance que i'ay de mon peu d'experience & de capacité ne me permet pas d'en auancer dauantage auât qu'il ait passé à l'examen des habiles gens, qui voudront nous obliger d'en prendre la peine; aprés quoy si l'on iuge que la chose merite d'estre continuée, nous essayrons de la pousser iusques où Dieu nous donnera la force de la conduire.

A PARIS, M. DC. XL.

---

paper and showed the influence of Desargues. A few years later Pascal invented a computing machine. At the age of twenty-five he decided to live the ascetic life of a Jansenist in the convent of Port-Royal, but continued to devote time to science and to literature. His treatise on the "arithmetical triangle" formed by the binomial coefficients and useful in probability appeared posthumously in 1664. We have already mentioned his work on integration and his speculations on the infinitesimal, which influenced Leibniz. He was also the first to establish a satisfying formulation of the principle of complete induction.[14]

Gérard Desargues was an architect from Lyons and the author of a book on perspective (1636). His pamphlet with the curious title *Brouillon projet d'une atteinte aux événements des rencontres d'un cone avec un plan* (1639)[15] contains in odd botanical language some of the fundamental conceptions of projective geometry, such as the points at infinity, involutions, and polarities. His "Desargues's theorem" on perspective triangles was published in 1648. These ideas did not show their full fertility until the nineteenth century.

7

A general method of differentiation and integration, derived in the full understanding that one process is the inverse of the other, could only be discovered by men who had mastered the geometrical method of the Greeks and of Cavalieri, as well as the algebraic method of Descartes and Wallis. Such men could have appeared only after 1660, and they actually did appear in Newton and Leibniz. Much has been written about the priority of the discovery, but it is now established that both men found their methods independently of each other. Newton had the calculus first (Newton in 1655–66; Leibniz in 1673–76), but Leibniz published it first (Leibniz 1684–86; Newton 1704–36). Leibniz's school was far more brilliant than Newton's school.

Isaac Newton was the son of a freehold farmer in Lincolnshire, England. He studied at Cambridge, became a fellow of Trinity College, and in 1669 was appointed Lucasian Professor, succeeding Isaac Barrow. Newton stayed at Cambridge until 1696, when he accepted the position of warden, and later of master, of the mint. In 1705 he was knighted by Queen Anne and became Sir Isaac. His tremendous authority is primarily based on his *Philosophiae naturalis principia mathematica* (1687), an enormous tome establishing mechanics on an axiomatic foundation and containing the law of gravitation—the law that brings the apple to the earth and keeps the moon moving around the earth. He showed by rigorous mathematical deduction how the empirically established laws of Kepler on planetary motion found their explication in the gravitational law of inverse squares and gave a dynamical explanation of many aspects of the motions of heavenly bodies and of the tides. He solved the two-body problem for spheres and laid the beginnings of a theory of the moon's motion. By solving the problem of the attraction of spheres he also laid the foundation of potential theory. His

[14]H. Freudenthal, *Arch. intern. des sciences*, Vol. 22 (1953), pp. 17–37; R. Rashed, *AHES*, Vol. 9 (1972), pp. 1–21.
[15]"Proposed draft of an attempt to deal with the events of the meeting of a cone with a plane."

axiomatic treatment postulated absolute space and absolute time.

The geometrical form of the demonstrations hardly shows that the author was in full possession of the calculus, which he called the "theory of fluxions." Newton discovered his general method during the years 1665–66 when he stayed at his birthplace in the country to escape from the plague that was infesting Cambridge. From this period also date his fundamental ideas on universal gravitation, as well as the law of the composition of light. "There are no other examples of achievement in the history of science to compare with that of Newton during those two golden years," remarks Professor More.[16]

Newton's discovery of "fluxions" was intimately connected with his study of infinite series through Wallis' *Arithmetica*. It brought him to extend the binomial theorem to fractional and negative exponents and thus to the discovery of the binomial series. This again helped him greatly in establishing his theory of fluxions to "all" functions, whether algebraic or transcendental. A "fluxion," expressed by a dot placed over a letter ("pricked letters"), was a finite value, a velocity; the letters without the dot represented "fluents."

Here is an example of the way in which Newton explained his method (*Method of Fluxions*, 1736): The variables of fluents are denoted by $v, x, y, z, \ldots$ "and the velocities by which every fluent is increased by its generating motion (which I may call *fluxions*, or simply velocities, or celerities), I shall represent by the same letters pointed, thus $\dot{v}, \dot{x}, \dot{y}, \dot{z}$." Newton's infinitesimals are called "moments of fluxions," which are represented by $\dot{v}o, \dot{x}o, \dot{y}o, \dot{z}o$, o being "an infinitely small quantity." Newton then proceeds:

Thus let any equation $x^3 - ax^2 + axy - y^3 = 0$ be given, and substitute $x + \dot{x}o$ for x, $y + \dot{y}o$ for y, and there will arise

$$x^3 + 3x^2\dot{x}o + 3x\dot{x}o\dot{x}o + \dot{x}^3o^3 - ax^2 - 2ax\dot{x}o - a\dot{x}o\dot{x}o + axy + ay\dot{x}o$$

$$+ ax\dot{y}o + ax\dot{y}o - y^3 - 3y^2\dot{y}o - 3y\dot{y}oyo - \dot{y}^3o^3 = 0.$$

Now, by supposition, $x^3 - ax^2 + axy - y^3 = 0$, which therefore, being expunged and the remaining terms being divided by o, there will remain

$$3x^2\dot{x} - 2ax\dot{x} + ay\dot{x} + ax\dot{y} - 3y^2\dot{y} + 3x\dot{x}\dot{x}o - a\dot{x}\dot{x}o$$

$$+ ax\dot{y}o - 3y\dot{y}\dot{y}o + \dot{x}^3oo - y^3oo = 0.$$

But whereas zero is supposed to be infinitely little, that it may represent the moments of quantities, the terms that are multiplied by it will be nothing in respect to the rest; I therefore reject them, and there remains

$$3x^2\dot{x} - 2ax\dot{x} + ay\dot{x} + ax\dot{y} - 3y^2\dot{y} = 0.$$

This example shows that Newton thought of his derivatives primarily as velocities, but it also shows that there was a certain vagueness in his mode of

---

[16]L. T. More, *Isaac Newton. A Biography* (New York, London, 1934), p. 41. See also Westfall's book, in the "Literature" section, below.

expression. Are the "o" symbols zeros? Are they infinitesimals? Or are they finite numbers? Newton tried to make his position clear by the theory of "prime and ultimate ratios," which he introduced in the *Principia* and which involved the conception of limit but in such a way that it was very hard to understand it.

> Those ultimate ratios with which quantities vanish are not truly the ratios of ultimate quantities, but limits toward which the ratios of quantities, decreasing without limit, do always converge, and to which they approach nearer than by any given difference, but never go beyond, nor in effect attain to, until the quantities have diminished in infinitum. (*Principio* I, Sect. I, last scholium.)

> Quantities, and the ratio of quantities, which in any finite time converge continually to equality, and before the end of that time approach nearer the one to the other than by any given difference, become ultimately equal. (*Principio* I, Sect. I, Lemma I.)

This was far from clear, and the difficulties which the understanding of Newton's theory of fluxions involved led to much confusion and severe criticism by Bishop Berkeley in 1734. The misunderstandings were not removed until the modern limit concept was well established.

Newton also wrote on conics and plane cubic curves. In the *Enumeratio linearum tertii ordinis* (1704) he gave a classification of plane cubic curves into seventy-two species, basing himself on his theorem that every cubic can be obtained from a "divergent parabola" $y^2 = ax^3 + bx^2 + cx + d$ by central projection from one plane upon another. This was the first important new result reached by the application of algebra to geometry, all previous work being simply the translation of Apollonius into algebraic language. Another contribution of Newton was his method of finding approximations to the roots of numerical equations, which he explained on the example $x^3 - 2x - 5 = 0$, which yields $x = 2.09455147$.

The difficulty in estimating Newton's influence on his contemporaries lies in the fact that he always hesitated to publish his discoveries. He first tested the law of universal gravitation in 1665–66, but did not announce it until he presented the manuscript of most of the *Principia* to the printer in 1686. His *Arithmetica universalis*, consisting of lectures on algebra delivered between 1673 and 1683, was published in 1707. His work on series, which dates from 1669, was announced in two letters to Oldenburg in 1676 and appeared in print in 1699. His quadrature of curves, of 1693, was not published until 1704; this was the first time that the theory of fluxions was fully placed before the world. His earlier *Method of Fluxions* only appeared in 1736, nine years after his death. The influence of the *Principia* was matched by that of the *Opticks* (1704 publication of a much older text). With the writings of Robert Boyle, it influenced the whole "experimental philosophy" of the eighteenth century.

8

Gottfried Wilhelm Leibniz was born in Leipzig and spent most of his life near the court of Hanover in the service of the dukes, one of whom became King of

ISAAC NEWTON (1642–1727)
(From a portrait by Godfrey Kneller.)

GOTTFRIED WILHELM LEIBNIZ (1646–1716)
(From a picture in the Uffizi Gallery, Florence.)

I.

NOVA METHODUS PRO MAXIMIS ET MINIMIS, ITEMQUE TANGENTIBUS, QUAE NEC FRACTAS NEC IRRATIONALES QUANTITATES MORATUR, ET SINGULARE PRO ILLIS CALCULI GENUS *).

Sit (fig. 111) axis AX, et curvae plures, ut VV, WW, YY, ZZ, quarum ordinatae ad axem normales, VX, WX, YX, ZX, quae vocentur respective v, w, y, x, et ipsa AX, abscissa ab axe, vocetur x. Tangentes sint VB, WC, YD, ZE, axi occurrentes respective in punctis B, C, D, E. Jam recta aliqua pro arbitrio assumta vocetur dx, et recta, quae sit ad dx, ut v (vel w, vel y, vel z) est ad XB (vel XC, vel XD, vel XE) vocetur dv (vel dw, vel dy, vel dz) sive differentia ipsarum v (vel ipsarum w, vel y, vel z). His positis, calculi regulae erunt tales.

Sit a quantitas data constans, erit da aequalis 0, et $\overline{dax}$ erit aequalis adx. Si sit y aequ. v (seu ordinata quaevis curvae YY aequalis cuivis ordinatae respondenti curvae VV) erit dy aequ. dv. Jam *Additio et Subtractio*: si sit z — y + w + x aequ. v, erit $dz - y + w + x$ seu dv aequ. dz — dy + dw + dx. *Multiplicatio*: $d\overline{xv}$ aequ. xdv + vdx, seu posito y aequ. xv, fiet dy aequ. xdv + vdx. In arbitrio enim est vel formulam, ut xv, vel compendio pro ea literam, ut y, adhibere. Notandum, et x et dx eodem modo in hoc calculo tractari, ut y et dy, vel aliam literam indeterminatam cum sua differentiali. Notandum etiam, non dari semper regressum a differentiali Aequatione, nisi cum quadam cautione, de quo alibi.

Porro *Divisio*: $d\dfrac{v}{y}$ vel (posito z aequ. $\dfrac{v}{y}$) dz aequ. $\dfrac{\pm vdy \mp ydv}{yy}$.

Quoad *Signa* hoc probe notandum, cum in calculo pro litera substituitur simpliciter ejus differentialis, servari quidem eadem signa, et pro + z scribi + dz, pro — z scribi — dz, ut ex addi-

---

*) Act. Erud. Lips. an. 1684.

England under the name of George I. He was even more catholic in his interests than the other great thinkers of his century; his philosophy embraced history, theology, linguistics, biology, geology, mathematics, diplomacy, and the art of inventing. He was one of the first after Pascal to invent a computing machine; he imagined steam engines, studied Chinese philosophy, and tried to promote the unity of Germany. The search for a universal method by which he could obtain knowledge, make inventions, and understand the essential unity of the universe was the mainspring of his life. The *scientia generalis* he tried to build had many aspects, and several of them led Leibniz to discoveries in mathematics. His search for a *characteristica generalis* led to permutations, combinations, and symbolic logic; his search for a *lingua universalis*, in which all errors of thought would appear as computational errors, led not only to symbolic logic but also to many innovations in mathematical notation. Leibniz was one of the greatest inventors of mathematical symbols. Few men have understood so well the unity of form and content. His invention of the calculus must be understood against this philosophical background; it was the result of his search for a *lingua universalis* of change and of motion in particular.

Leibniz found his new calculus between 1673 and 1676 in Paris under the personal influence of Huygens and by the study of Descartes and Pascal. He was stimulated by his knowledge that Newton was reported to be in the possession of such a method. Where Newton's approach was primarily kinematical, Leibniz's was geometrical; he thought in terms of the "characteristic triangle" (dx, dy, ds), which had already appeared in several other writings, notably in Pascal and in Barrow's *Geometrical Lectures* of 1670.[17] The first publication of Leibniz's form of calculus occurred in 1684 in a six-page article in the *Acta eruditorum*, a mathematical periodical which he had helped to found in 1682. The paper had the characteristic title, *Nova methodus pro maximis et minimis, itemque tangentibus, quae nec fractas nec irrationales quantitates moratur, et singulare pro illis calculi genus.*[18] It was a barren and obscure account, but it contained our symbols dx, dy and the rules of differentiation, including $d(uv) = udv + vdu$ and the differential for the quotient, with the condition $dy = 0$ for extreme values and $d^2y = 0$ for points of inflection. This paper was followed in 1686 by another (written in the form of a book review) with the rules of the integral calculus, containing the $\int$ symbol. It expressed the equation of the cycloid as

$$y = \sqrt{2x - x^2} + \int \frac{dx}{\sqrt{2x - x^2}}.$$

An extremely fertile period of mathematical productivity began with the publication of these papers. Leibniz was joined after 1687 by the Bernoulli brothers, who eagerly absorbed his methods. Before 1700 these men had found most of our

---

[17]The term *triangulum characteristicum* seems to have first been used by Leibniz, who found it by reading Pascal's *Traité des sinus du quart de cercle*, part of his Dettonville letters (1658). It had already occurred in Snellius' *Tiphys Batavus* (1624), pp. 22–25.

[18]"A new method for maxima and minima, as well as tangents, which is not obstructed by fractional and irrational quantities, and a curious type of calculus for it."

undergraduate calculus, together with important sections of more advanced fields, including the solution of some problems in the calculus of variations. By 1696 the first textbook on calculus appeared, the *Analyse des infiniment petits*, written by the Marquis de l'Hospital under the strong influence of Johann Bernoulli, who for a while had tutored him. This book, for a long time unique in its field, contains the so-called "rule of l'Hospital" for finding the limiting value of a fraction whose two terms both tend toward zero.[19]

Our notation of the calculus is due to Leibniz, and even the names *calculus differentialis* and *calculus integralis*.[20] Because of his influence, the sign = is used for equality and the × for multiplication. The terms "function" and "coordinates" are due to Leibniz, as well as the playful term "osculating." The series

$$\frac{\pi}{4} = \frac{1}{1} - \frac{1}{3} + \frac{1}{5} - \frac{1}{7} + \cdots$$

and

$$\tan^{-1} x = x - \frac{x^3}{3} + \frac{x^5}{5} - \cdots$$

are named after Leibniz, though the priority of the discovery was not his. This seems to go to James Gregory, a mathematician from an intellectually distinguished Scotch family, who after having spent some years (1664–68) abroad, mostly at Padua, taught at St. Andrews from 1668 until shortly before he died at thirty-seven years of age. His letters, and the three books he wrote while in Italy, among them the *Exercitationes Geometricae* (1668), show his great originality in dealing with infinite processes. He found the binomial series (1670), and in 1671 even arrived at Taylor's series. If he had lived he might have ranked with Newton and Leibniz as an inventor of the calculus.

Leibniz's explanation of the foundations of the calculus suffered from the same vagueness as Newton's. Sometimes his dx, dy were finite quantities, sometimes quantities less than any assignable quantity and yet not zero. In the absence of rigorous definitions he presented analogies pointing to the relation of the radius of the earth to the distance of the fixed stars. He varied his modes of approach to questions concerning the infinite; in one of his letters (to Foucher, 1693) he accepted the existence of the actually infinite to overcome Zeno's difficulties and praised Grégoire de Saint-Vincent, who had computed the place where Achilles meets the tortoise. And just as Newton's vagueness provoked the criticism of Berkeley, so Leibniz's vagueness provoked the opposition of Bernard Nieuwentijt, burgomaster of Purmerend near Amsterdam (1694). Both Berkeley's and Nieuwentijt's criticism had their justification, but they were entirely negative. They were unable to supply a rigorous foundation to the calculus, but inspired further constructive work.

---

[19]This rule was communicated to L'Hospital by Johann Bernoulli in a letter, which only recently has come to light: J. Bernoulli, *Briefwechsel* I (Basel, 1955). [See on this D. J. Struik, *Mathematics Teacher*, Vol. 56 (1963), 257–60.]

[20]Leibniz suggested the name *calculus summatorius* first, but in 1696 Leibniz and Johann Bernoulli agreed on the name *calculus integralis*. Modern analysis has returned to Leibniz's early terminology. See further: F. Cajori, "Leibniz, the Master Builder of Mathematical Notations," *Isis*, Vol. 7 (1925), pp. 412–29.

Literature

The collected works of Kepler, Galileo, Descartes, Pascal, Huygens, Fermat, and Newton are available in modern editions; those of Mersenne and Leibniz, only in part:

[Whiteside, D. T., and Hoskins, M. A., eds.] *Mathematical Papers of Isaac Newton.* 8 vols., Cambridge, 1976–81.

[———.] *The Mathematical Works of Isaac Newton.* 2 vols., New York and London, 1964–67, with facsimile reproductions and introductions by Turnbull, H. W., and Scott.

[Koyré, A., Cohen, I. B., and Whitman, A., eds.] *Isaac Newton's Philosophiae Naturalis Principia Mathematica, Third Edition (1726) with Variant Readings.* 2 vols., Cambridge, Mass., and Cambridge, England, 1972.

Newton, I. *The Mathematical Principles of Natural Philosophy.* Trans. by A. Motte, 1728. Ed. and introd. by I. B. Cohen. 2 vols., London, 1968.

Turnbull, H. W. *The Mathematical Discoveries of Newton.* Glasgow, 1934.

More, L. T. *Isaac Newton. A Biography.* New York and London, 1934.

Westfall, R. S. *Never at Rest. A Biography of I. Newton.* New York, etc., 1981.

Sheynin, O. B. "Newton and the Classical Theory of Probability." *AHES*, Vol. 7 (1971), pp. 217–56.

See on Newton further the extensive article by I. B. Cohen in *DSB*, Vol. 10 (1974), pp. 42–103, which includes a report on Newton in the USSR.

The Deutsche Akademie der Wissenschaften in Berlin is publishing the *Sämtliche Schriften und Briefe* (1923) of Leibniz, but has not reached the mathematical works. These remain in:

[C. I. Gerhardt, ed.] *G. W. Leibniz' mathematische Schriften.* 7 vols., Berlin and Halle, 1849–63; republished Hildesheim, 1962. ("Register" [Index] by J. E. Hofmann, Hildesheim, 1977.)

Child, J. M. *The Early Mathematical Manuscripts of Leibniz.* Chicago, 1920 (translated from the Latin originals).

Couturat, L. *La Logique de Leibniz.* Paris, 1901.

———. *Opuscules et fragments inédits de Leibniz.* Paris, 1906; republished Hildesheim, 1961.

Knobloch, E. *Die mathematischen Studien von G. W. Leibniz zur Kombinatorik.* 2 vols., Wiesbaden, 1973, 1976.

Leibniz à Paris, Studia Leibnitziana Supplementa 18, Vol. 2 (1978), 371 pp. (thirteen essays).

Hofmann, J. E. *Die Entwicklungsgeschichte der Leibnizchen Mathematik während des Aufenthaltes in Paris* (1672–1676). Munich, 1949. See, regarding these two books, *HM*, Vol. 9 (1982), pp. 109–23. (Other studies by Hofmann on seventeenth-century mathematicians include those on N. Mercator [*Deut. Math.*, Vol. 3 (1939); Vol. 5 (1940)]; Grégoire de Saint-Vincent [*Abh. Preuss. Akad. Wiss., Math. Naturw.*

Klasse, No. 31 (1941)]; Fermat [*Ibid.*, No. 7 (1944)]; and on the priority struggle between Newton and Leibniz [*Ibid.*, No. 2 (1944)]. Cf. the bibliographical data in his *Geschichte der Mathematik*, 3 vols., Berlin, 1953–57. Also, *Frans van Schooten der Jüngere*. Wiesbaden, 1962.)
See on Leibniz further the article by J. E. Hofmann, *DSB*, Vol. 8 (1973), pp. 149–68.
Aus den Frühzeit der Infinitesimalmethoden, *AHES*, Vol. 2 (1964), pp. 271–343.
Full bibliography by C. J. Scriba in *Mitt. aus dem math. Seminar Giessen*, 1971, pp. 51–73; and *HM*, Vol. 2 (1975), pp. 148–52.

For the discovery of the calculus, see:

Boyer, C. B. *The History of the Calculus*. New York, 1949. Dover reprint, 1959. (Contains a large bibliography.)

On the historical-technical background, see:

Grossman, H. "Die gesellschaftlichen Grundlagen der mechanistischen Philosophie und die Manufaktur." *Zeitschrift für Sozialforschung*, Vol. 4 (1935), pp. 161–231.
Merton, R. K. "Science, Technology and Society in the Seventeenth Century." *Osiris*, Vol. 4 (1938), pp. 360–62; also *Science and Society*, Vol. 3 (1939), pp. 3–27.

On leading mathematicians, see:

Cajori, F. *William Oughtred*. Chicago and London, 1916.
Scott, J. F. *The Mathematical Works of John Wallis D. D., F.R.S.* London, 1938. Reprinted: New York, 1981.
Prag, A. "John Wallis, Zur Ideengeschichte der Mathematik im 17. Jahrhundert." *Quellen und Studien*, Vol. 1 (1930), pp. 381–412. (Also see T. P. Nunn, *Math. Gazette*, Vol. 5 [1910–11].)
Hessen, B. "The Social and Economic Roots of Newton's *Principia*." In *Science at the Crossroads*. London, 1934.
Scriba, C. J. "Studien zur Mathematik des John Wallis (1616–1703)." *Boethius*, Vols. 6, 7. (Wiesbaden, 1966.)
Mahoney, M. S. *The Mathematical Career of Pierre Fermat*. Princeton, N.J., 1970.
Barrow, I. *Geometrical Lectures*. Chicago, 1916. (Trans. and ed. by J. M. Child.)
Bell, A. E. *Christian Huygens and the Development of Science in the Seventeenth Century*. London, 1947.
Johann Kepler. A Tercentenary Commemoration of His Life and Works. Baltimore, 1931.
Milhaud, G. *Descartes savant*. Paris, 1921.
Bosmans, H. (see Chap. V) has papers on: Tacquet (*Isis*, Vol. 9 [1927–28], pp. 66–83), Stevin (*Mathesis*, Vol. 37 [1923]; *Ann. Soc. Sc. Bruxelles*, Vol. 37 [1913], pp. 171–99; *Biographie nationale de Belgique*), Della Faille (*Mathesis*, Vol. 41 [1927], pp. 5–11), De Saint-Vincent (*Mathesis*, Vol. 38 [1924], pp. 250–56).
Toeplitz, O. *Die Entwicklung der Infinitesimalrechnung I*. Berlin, 1949.
Taton, R. *L'oeuvre mathématique de G. Desargues*. Paris, 1951.
James Gregory, A Tercentenary Memorial, H. W. Turnbull, ed. London, 1939. [Cf. M. Dehn and E. D. Hellinger, *Amer. Math. Monthly*, Vol. 50 (1943), pp. 149–63.]
Scriba, C. J. *James Gregorys frühe Schriften zur Infinitesimalrechnung*. Mitt. aus dem math. Seminar Giessen, 55 (1957), 80 pp.

Haas, K. "Die mathematischen Arbeiten von Johannes Hudde." *Centaurus*, Vol. 4 (1956), pp. 235–84.

Fellman, E. A. "Die mathematischen Werke von Honoratius Fabri." *Physis*, Vol. 1 (1959), pp. 1–54.

Whiteside, D. T. "Patterns of Mathematical Thought in the Later Seventeenth Century." *AHES*, Vol. 1 (1961), pp. 179–388.

Tannery, Paul. "Notions Historiques." In *Notions de mathématiques*, J. Tannery, ed. Paris, 1903, pp. 324–48.

Montel, P. *Pascal mathématicien*. Paris, 1951.

Fleckenstein, J. O. *Die Prioritätsstreit zwischen Leibniz und Newton*. Basel and Stuttgart, 1956.

Lohne, J. A. "Thomas Harriot als Mathematiker," *Centaurus*, Vol. 11 (1965), pp. 19–45; also *DSB*, Vol. 6 (1972), pp. 124–29.

Bos, H. J. M. "Differentials, Higher-order Differentials and the Derivative in the Leibnizian Calculus," *AHES*, Vol. 14 (1974), pp. 1–90.

[Smith, D. E., and Latham, H., eds.] *The Geometry of René Descartes*. Facsimile text plus English translation, Chicago 1925, reprint New York, 1959.

Hall, A. R. *Philosophers at War. The Quarrel Between Newton and Leibniz*. Cambridge, 1900.

Struik, D. J. *The Land of Stevin and Huygens*. Dordrecht, 1981. Trans. from the Dutch: Amsterdam, 1958; Nijmegen, 1979.

Taton, R. "L'oeuvre de Pascal en Géométrie Projective." *Rev. Hist. Science Appl.*, Vol. 15 (1962), pp. 197–252.

Mulcrone, T. E. "A Catalog of Jesuit Mathematicians." *Bull. Amer. Assoc. Jesuit Scientists, Eastern States Division*, Vol. 41 (1964), pp. 83–90.

Auger, L. *Un savant méconnu, Giles Personne de Roberval, 1602–1675*. Paris, 1962.

Dugas, R. *La mécanique au XVIIe siècle*. Neuchâtel, 1954.

Baron, M. E. *The Origins of the Infinitesimal Calculus*. New York, 1969.

CHAPTER VII

The Eighteenth Century

1

Mathematical productivity in the eighteenth century concentrated on the calculus and its application to mechanics. The major figures can be arranged in a kind of pedigree to indicate their intellectual kinship:

Leibniz—(1646–1716)
The brothers Bernoulli: Jakob (1654–1705), Johann (1667–1748)
Euler—(1707–1783)
Lagrange—(1736–1813)
Laplace—(1749–1827).

Closely related to the work of these men was the activity of a group of French mathematicians, notably Clairaut, d'Alembert, and Maupertuis, who were again connected with the philosophers of the Enlightenment. To them must be added the Swiss mathematicians Lambert and Daniel Bernoulli. Scientific activity usually centered around academies, of which those at Paris, Berlin, and St. Petersburg were outstanding. University teaching played a minor role or no role at all. It was a period in which some of the leading European countries were ruled by what have euphemistically been called enlightened despots: Frederick the Great, Catherine the Great, to whom we may also add Louis XV and Louis XVI. Part of these despots' claim to glory was their delight in surrounding themselves with learned men. This delight was a type of intellectual snobbery, tempered by some understanding of the important role which natural science and applied mathematics were taking in improving manufactures and increasing the efficiency of the military. It is said, for instance, that the excellence of the French navy was due to the fact that in the construction of frigates and of ships of the line the master shipbuilders were partly led by mathematical theory. Euler's works abound in applications to questions of importance to the army and navy. Astronomy continued to play its outstanding role as foster-mother to mathematical research under royal and imperial protection.

2

Basel in Switzerland, a free empire city since 1263, had long been a center of learning. In the days of Erasmus its university was already a great center. The arts and sciences flourished in Basel, as in the cities of Holland, under the rule of a merchant patriciate. To this Basel patriciate belonged the merchant family of the Bernoullis, who had come via Amsterdam from Antwerp in the previous century after the latter city had been conquered by the Spanish. From the late seventeenth century to the present time this family in every generation has produced scientists. Indeed it is difficult to find in the whole history of science a family with a more distinguished record.

This record begins with two mathematicians, Jakob (James, Jacques) and Johann (John, Jean) Bernoulli. Jakob studied theology, Johann studied medicine; but when Leibniz's papers in the *Acta eruditorum* appeared both men decided to become mathematicians. They became the first important pupils of Leibniz. In 1687 Jakob accepted the chair of mathematics at Basel University, where he taught until his death in 1705. In 1695 Johann became professor at Groningen; upon his older brother's death he succeeded him in the chair at Basel, where he remained for forty-three more years.

Jakob began his correspondence with Leibniz in 1687. Then, in a constant exchange of ideas with Leibniz and with each other—often with bitter rivalry between them—the two brothers began to discover the treasures contained in Leibniz's pioneering venture. The list of their results is long and includes not only much of the material now contained in our elementary texts on differential and integral calculus, but also material on the integration of many ordinary differential equations. Among Jakob's contributions are the use of polar coordinates, the study of the catenary (already discussed by Huygens and others), the lemniscate (1694), and the logarithmic spiral. In 1690 he found the so-called isochrone, proposed by Leibniz in 1687 as the curve along which a body falls with uniform velocity; it appeared to be a semicubic parabola. Jakob also discussed isoperimetric figures (1701), which led to a problem in the calculus of variations. The logarithmic spiral, which has a way of reproducing itself under various transformations (its evolute is a logarithmic spiral and so are both the pedal curve and the caustic with respect to the pole), was such a delight to Jakob that he had it engraved on his tombstone with the inscription *eadem mutata resurgo*.[1]

Jakob Bernoulli was also one of the early students of the theory of probabilities, on which subject he wrote the *Ars conjectandi*, published posthumously in 1713. In the first part of this book Huygens' tract on games of chance is reprinted; the other parts deal with permutations and combinations and come to a climax in the "theorem of Bernoulli" on binomial distributions. "Bernoulli's numbers" appear in this book in a discussion of Pascal's triangle.

3

Johann Bernoulli, twelve and a half years younger than his brother, lived to be eighty years old, working himself up, from his position at Basel, to a kind of—

---

[1] "I arise the same though changed." The spiral on the gravestone, however, looks more like an Archimedean spiral.

rather quarrelsome—elder statesman in the mathematical world, proud, however, of his unique pupil, Leonhard Euler. His work was closely related to that of his older brother, and it is not always easy to distinguish between the results of the two men. Both Jakob and Johann are often considered the inventors of the calculus of variations because of their contribution to the problem of the brachistochrone. This is the curve of quickest descent for a mass point moving between two points in a gravitational field, a curve studied by Leibniz and the Bernoullis in 1697 and the following years. At this time they found the equation of the geodesics on a surface.[2] The answer to the problem of the brachistochrone is the cycloid. This curve also solves the problem of the tautochrone, the curve along which a mass point in a gravitational field reaches the lowest point in a time independent of its starting point. Huygens discovered this property of the cycloid and used it in constructing tautochronous pendulum clocks (1673) in which the period is independent of the amplitude.

Among the other Bernoullis who have influenced the course of mathematics, Johann's son, Daniel, stands out. After a stay at St. Petersburg (1725–33), after 1727 together with Euler, he became professor at Basel, where he lived to a ripe old age. In his early years, he collaborated with his cousin Nikolaus (I) on that problem in probability known as the problem (or more dramatically as the "paradox") of St. Petersburg, since the work was published in that city.[3] His prolific activity was centered on applications ("mixed mathematics" it was called): astronomy, physics, physiology, and hydrodynamics, a term due to him. The *Hydrodynamica* (1736) contains Bernoulli's law on hydraulic pressure and in Chapter 16 the principles of the kinetic theory of gases. It inspired Euler in his fundamental work on the dynamics of fluids. With Euler and d'Alembert, Daniel tackled the problem of the vibrating string, first introduced by Brook Taylor (1715). This led to the theory of partial differential equations, not without controversy, especially on the role of trigonometric series, here introduced for the first time.

Ordinary differential equations had been the domain of Daniel's father and uncle. One other son of Johann, Johann (II), and two children of this Johann, Johann (III) and Jakob (II), also had distinguished careers in mathematics. And they were not the only members of this family with academic careers.[4]

---

[2]Newton in a scholium of the *Principia* (II, Prop. 35) had already discussed the solid of revolution moving in a liquid with least resistance. He published no proof of his contention.

[3]*Comm. Acad. Scient. Imp. Petropolitanae*, Vol. 5 (1730/31), publ. 1738. It is after this publication that the problem is named, and not, as is sometimes believed, after the stay of the other Nikolaus (II), brother of Daniel, in St. Petersburg.

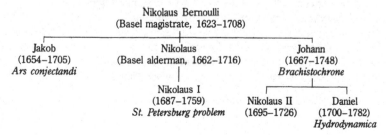

Nikolaus Bernoulli
(Basel magistrate, 1623–1708)

Jakob
(1654–1705)
Ars conjectandi

Nikolaus
(Basel alderman, 1662–1716)

Johann
(1667–1748)
Brachistochrone

Nikolaus I
(1687–1759)
St. Petersburg problem

Nikolaus II
(1695–1726)

Daniel
(1700–1782)
Hydrodynamica

4

Also from Basel came the most productive mathematician of the eighteenth century—if not of all times—Leonhard Euler. His father studied mathematics under Jakob Bernoulli and Leonhard studied it under Johann. When in 1725 Johann's son Nikolaus traveled to St. Petersburg, young Euler followed him and stayed at the Academy until 1741. From 1741 to 1766 Euler was at the Berlin Academy under the special tutelage of Frederick the Great; from 1766 to 1783 he was again at St. Petersburg, now under the aegis of the Empress Catherine. He married twice and had thirteen children. The life of this eighteenth-century academician was almost exclusively devoted to work in the different fields of pure and applied mathematics. Although he lost one eye in 1735 and the other eye in 1766, nothing could interrupt his enormous productivity. The blind Euler, aided by a phenomenal memory, continued to dictate his discoveries. During his life 560 books and papers appeared; at his death he left many manuscripts, which were published by the St. Petersburg Academy during the next forty-seven years. (This brought the number of his works to 771, but research by Gustav Eneström has completed the list to make 856, plus 31 works by his eldest son, Johann Albrecht, written under his father's supervision—all this apart from an extensive correspondence.)

Euler made signal contributions in every field of mathematics which existed in his day. He published his results not only in articles of varied length, but also in an impressive number of large textbooks which ordered and codified the material assembled during the ages. In several fields Euler's presentation has been almost final. An example is our present trigonometry with its conception of trigonometric values as ratios and its useful notation, which dates from Euler's *Introductio in analysin infinitorum* (1748). The tremendous prestige of his textbooks settled forever many moot questions of notation in algebra and calculus; Lagrange, Laplace, and Gauss knew and followed Euler in all their works.

The *Introductio* of 1748 covers in its two volumes a wide variety of subjects. It contains an exposition of infinite series including those for e^x, sin x, and cos x, and presents the relation $e^{ix} = \cos x + i \sin x$ (already discovered by Johann Bernoulli and others in different forms). Curves and surfaces are so freely investigated with the aid of their equations that we may consider the *Introductio* the first text of analytic geometry. We may also find here an algebraic theory of elimination. To the most exciting parts of this book belong the chapter on the Zeta function and its relation to the prime number theory, as well as the chapter on *partitio numerorum*.[5]

Another great and rich textbook was Euler's *Institutiones calculi differentialis* (1755), followed by three volumes of *Institutiones calculi integralis* (1768–74). Here we find not only our elementary differential and integral calculus but also a theory of differential equations, Taylor's theorem with many applications, Euler's "summation" formula, and the Eulerian integrals Γ and B. The section on differential equations with its distinction between "linear," "exact," and "homoge-

---

[5]See the preface to the *Introductio* by A. Speiser: Euler, *Opera omnia*, 1st ser., Vol. 9 (1945).

neous" equations is still the model of our elementary texts on this subject. Euler also sharpened the concept of *function*.

Euler's *Mechanica, sive motus scientia analytice exposita* (1736) was the first textbook in which Newton's dynamics of the mass point was developed with analytical methods. It was followed by the *Theoria motus corporum solidorum seu rigidorum* (1765), in which the mechanics of solid bodies was similarly treated. This textbook contains the "Eulerian" equations for a body rotating about a point. These books establish rational mechanics in the form that we know from our textbooks, while its inspiration, the *Principia* of Newton, is left to the specialized historian of science. The *Vollständige Anleitung zur Algebra* (1770)—written in German and dictated to a servant because of Euler's blindness—has been the model of many later texts on algebra. It leads up to the theory of cubic and biquadratic equations and ends with a chapter on indeterminate equations, in which we find the proofs that $x^n + y^n = z^n$ is impossible for integers $x, y, z, n = 3$ and $n = 4$ (see Chap. VI, Sec. 6, above, on Fermat's problem).

In 1744 appeared Euler's *Methodus inveniendi lineas curvas maximi minimive proprietate gaudentes*. This was the first exposition of the calculus of variations; it contained "Euler's equations" with many applications, including the discovery that catenoid and right helicoid are minimal surfaces. Many other results of Euler can be found in his smaller papers, which contain many a gem, little known even today. To the better-known discoveries belong the theorem connecting the number of vertices (V), edges (E), and faces (F) of a closed polyhedron ($V + F - E = 2$);[6] the line of Euler in the triangle; the curves of constant width (Euler called them orbiform curves); and the constant of Euler

$$\lim_{n \to \infty} \left(\frac{1}{1} + \frac{1}{2} + \cdots + \frac{1}{n} - \log n \right) = .5772156 \ldots$$

Several papers are devoted to mathematical recreations (the seven bridges of Königsberg, the knight's move in chess). Euler's contributions to the theory of numbers alone would have given him a niche in the hall of fame; to his discoveries in this field belongs the law of reciprocity for quadratic residues.

A large amount of Euler's activity was devoted to astronomy, where the lunar theory, important both as a section of the three-body problem and in the solution of the problem of finding longitude, received his special attention. The *Theoria motus planetarum et cometarum* (1774) is a treatise on celestial mechanics. Related to this work was Euler's study of the attraction of ellipsoids (1738).

There are books by Euler on hydraulics, on ship construction, on artillery. In 1769–71 there appeared three volumes of a *Dioptrica* with a theory of the passage of rays through a system of lenses. In 1739 appeared his new theory of music, of which it has been said that it was too musical for mathematicians and too mathematical for musicians. Euler's philosophical exposition of the most important problems of natural science in his *Letters to a German Princess* (1760–61, in French) remained a model of popularization.

---

[6] Already known to Descartes, *Oeuvres* (C. Adam and P. Tannery, Eds.), Vol. X, pp. 257–76.

Secantes autem et cosecantes ex tangentibus per solam subtractionem inveniuntur; est enim

$$\text{cosec. } z = \text{cot. } \frac{1}{2} z - \text{cot. } z$$

et hinc

$$\text{sec. } z = \text{cot. } \left(45° - \frac{1}{2} z\right) - \text{tang. } z.$$

Ex his ergo luculenter perspicitur, quomodo canones sinuum construi potuerint.

138. Ponatur denuo in formulis § 133 arcus z infinite parvus et sit n numerus infinite magnus i, ut iz obtineat valorem finitum v. Erit ergo $nz = v$ et $z = \frac{v}{i}$, unde sin. $z = \frac{v}{i}$ et cos. $z = 1$; his substitutis fit

$$\text{cos. } v = \frac{\left(1 + \frac{v\sqrt{-1}}{i}\right)^i + \left(1 - \frac{v\sqrt{-1}}{i}\right)^i}{2}$$

atque

$$\text{sin. } v = \frac{\left(1 + \frac{v\sqrt{-1}}{i}\right)^i - \left(1 - \frac{v\sqrt{-1}}{i}\right)^i}{2\sqrt{-1}}.$$

In capite autem praecedente vidimus esse

$$\left(1 + \frac{z}{i}\right)^i = e^z$$

denotante e basin logarithmorum hyperbolicorum; scripto ergo pro z partim $+ v\sqrt{-1}$ partim $- v\sqrt{-1}$ erit

$$\text{cos. } v = \frac{e^{+v\sqrt{-1}} + e^{-v\sqrt{-1}}}{2}$$

et

$$\text{sin. } v = \frac{e^{+v\sqrt{-1}} - e^{-v\sqrt{-1}}}{2\sqrt{-1}}.$$

Ex quibus intelligitur, quomodo quantitates exponentiales imaginariae ad sinus et cosinus arcuum realium reducantur.[1] Erit vero

---

1) Has celeberrimas formulas, quas ab inventore *Formulas Eulerianas* nominare solemus, Eulerus distincte primum exposuit in Commentatione 61 (indicis Enestroemiani): *De summis*

THE PAGES IN EULER'S "INTRODUCTIO" WHERE
$e^{ix} = \cos x + i \sin x$ IS INTRODUCED.
(Taken from a later reproduction of the 1748 text. Euler's formula was published in 1743, and even earlier, in letters to Goldbach of 1741 and 1742.)

et
$$e^{+v\sqrt{-1}} = \cos. v + \sqrt{-1} \cdot \sin. v$$

$$e^{-v\sqrt{-1}} = \cos. v - \sqrt{-1} \cdot \sin. v.$$

139. Sit iam in ' iisdem formulis § 133 n numerus infinite parvus seu $n = \frac{1}{i}$ existente i numero infinite magno; erit

$$\cos. nz = \cos. \frac{z}{i} = 1 \quad \text{et} \quad \sin. nz = \sin. \frac{z}{i} = \frac{z}{i};$$

arcus enim evanescentis $\frac{z}{i}$ sinus est ipsi aequalis, cosinus vero $= 1$. His positis habebitur

$$1 = \frac{(\cos. z + \sqrt{-1} \cdot \sin. z)^{\frac{1}{i}} + (\cos. z - \sqrt{-1} \cdot \sin. z)^{\frac{1}{i}}}{2}$$

et

$$\frac{z}{i} = \frac{(\cos. z + \sqrt{-1} \cdot \sin. z)^{\frac{1}{i}} - (\cos. z - \sqrt{-1} \cdot \sin. z)^{\frac{1}{i}}}{2\sqrt{-1}}.$$

Sumendis autem logarithmis hyperbolicis supra (§ 125) ostendimus esse

$$l(1 + x) = i(1 + x)^{\frac{1}{i}} - i \quad \text{seu} \quad y^{\frac{1}{i}} = 1 + \frac{1}{i} ly$$

posito y loco $1 + x$. Nunc igitur posito loco y partim $\cos. z + \sqrt{-1} \cdot \sin. z$ partim $\cos. z - \sqrt{-1} \cdot \sin. z$ prodibit

serierum reciprocarum ex potestatibus numerorum naturalium ortarum, Miscellanea Berolin. 7, 1743, p. 172; LEONHARDI EULERI Opera omnia, series I, vol. 14. Iam antea quidem cum amico CHR. GOLDBACH (1690—1764) formulas huc pertinentes, partim speciales partim generaliores, communicaverat. Sic in epistola d. 9. Dec. 1741 scripta invenitur haec formula

$$\frac{2^{+\sqrt{-1}} + 2^{-\sqrt{-1}}}{2} = \text{Cos. Arc. } l\, 2$$

et in epistola d. 8. Maii 1742 scripta haec

$$a^{p\sqrt{-1}} + a^{-p\sqrt{-1}} = 2 \text{ Cos. Arc. } pla.$$

Vide *Correspondance math. et phys. publiée par P. H. Fuss*, St.-Pétersbourg 1843, t. I, p. 110 et 123; LEONHARDI EULERI Opera omnia, series III. Confer etiam Commentationem 170 nota 1 p. 35 laudatam, imprimis § 90 et 91.   A. K.

The enormous fertility of Euler has been a continuous source of surprise and admiration for everyone who has attempted to study his work, a task not so difficult as it seems, since Euler's Latin is very simple and his notation is almost modern—or perhaps we should better say that our notation is almost Euler's! A long list can be made of the known discoveries for which Euler possesses priority, and another of ideas which are still worth elaborating. Great mathematicians have always appreciated their indebtedness to Euler. *"Lisez Euler,"* Laplace used to say to younger mathematicians, *"lisez Euler, c'est notre maître à tous."* And Gauss, more ponderously, expressed himself: "The study of Euler's works will remain the best school for the different fields of mathematics and nothing else can replace it." Riemann knew Euler's works well and some of his most profound writings have an Eulerian touch. Publishers might do worse than offer translations of some of Euler's works together with modern commentaries.

5

It is instructive to point out not only some of Euler's contributions to science but also some of his weaknesses. Infinite processes were still carelessly handled in the eighteenth century and much of the work of the leading mathematicians of that period impresses us as wildly enthusiastic experimentation. There was experimentation with infinite series, with infinite products, with integration, with the use of symbols such as 0, ∞, $\sqrt{-1}$. If many of Euler's conclusions can be accepted today, there are others concerning which we have reservations. We accept, for instance, Euler's statement that $\log n$ has an infinity of values which are all complex numbers, except when n is positive, when one of the values is real. Euler came to this conclusion in a letter to d'Alembert (1747), who had claimed that $\log(-1) = 0$. But we cannot always follow Euler when he writes that $1 - 3 + 5 - 7 + \ldots = 0$, or when he concludes from

$$n + n^2 + \cdots = \frac{n}{1 - n}$$

and

$$1 + \frac{1}{n} + \frac{1}{n^2} + \cdots = \frac{n}{n - 1}$$

that

$$\cdots + \frac{1}{n^2} + \frac{1}{n} + 1 + n + n^2 + \cdots = 0.$$

Yet we must be careful not to criticize Euler too hastily for his way of manipulating divergent series; he simply did not always use some of our present tests of convergence or divergence as a criterion for the validity of his series. Much of his supposedly indiscriminate work with series has been given a strictly rigorous sense by modern mathematicians.[7]

We cannot, however, be enthusiastic about Euler's way of basing the calculus on the introduction of zeros of different orders. An infinitesimally small quantity, wrote Euler in his *Differential Calculus* of 1755, is truly zero, $a \pm ndx = a$,[8] $dx \pm (dx)^{n+1} = dx$, and $a\sqrt{dx} + Cdx = a\sqrt{dx}$.

> Therefore there exist infinite orders of infinitely small quantities, which, though they all $= 0$, still have to be well distinguished among themselves, if we look at their mutual relation, which is explained by a geometrical ratio.[9]

The whole question of the foundation of the calculus remained a subject of debate, and so did all questions relating to infinite processes. The "mystical" period in the foundation of the calculus (to use a term suggested by Karl Marx) itself provoked a mysticism which occasionally went far beyond that of the founding fathers. Guido Grandi, a priest and professor at Pisa known for his study of rosaces ($r = \sin n\theta$) and other curves which resemble flowers,[10] considered the formula

$$\tfrac{1}{2} = 1 - 1 + 1 - 1 + 1 - \cdots$$

$$= (1 - 1) + (1 - 1) + (1 - 1) + \cdots$$

$$= 0 + 0 + 0 + \cdots$$

as the symbol for Creation from Nothing. He obtained the result 1/2 by considering the case of a father who bequeaths a gem to his two sons, who each may keep the bauble one year in alternation. It then belongs to each son for one half.

Euler's foundation of the calculus may have had its weakness, but he expressed his point of view without vagueness. D'Alembert, in some articles of the *Encyclopédie*, attempted to find this foundation by other means. Newton had used the term "prime and ultimate ratio" for the "fluxion," as the first or last ratio of two quantities just springing into being. D'Alembert replaced this notion by the conception of a "limit." One quantity he called the limit of another when the

---

[7]"We should not—as has often been done, especially towards the end of the nineteenth-century—speak of lack of rigor when we deal with methods of previous mathematicians that we can handle in a satisfactory way with the aid of techniques that we have acquired in the course of time. We should only speak of lack of rigor when some result has been obtained by means of a reasoning that cannot be logically maintained." ("Introduction to Euler," *Opera omnia*, C. Carathéodory, ed., 1st ser., Vol. 24 [1952], p. xvi.) On Euler and divergent series, see also G. H. Hardy, *Divergent Series* (Oxford, 1949).

[8]This formula reminds us of a statement ascribed to Zeno by Simplicius: "That which, being added to another, does not make it greater, and being taken away from another, does not make it less, is nothing."

[9]Euler, *Opera Omnia*, 1st ser., Vol. 10, p. 72. The reaction of most mathematicians has been—and still is—that even the good Euler sometimes slept. Professor A. P. Juškevič has pointed out that there may also be another side to the question: *Euler und Lagrange über die Grundlagen der Analysis*, in *Euler Sammelband zum 250. Geburtstage* (Berlin, 1959), pp. 224–44.

[10]L. Tenca, "Guido Grandi," *Physis* 2 (1960), pp. 84–89.

second approaches the first nearer than by any given quantity. "The differentiation of equations consists simply in finding the limits of the ratio of finite differences of two variables included in the equation" (see Section 7 of this chapter). This was a great step ahead, as was d'Alembert's conception of infinites of different orders. However, his contemporaries were not easily convinced of the importance of the new step, and when d'Alembert said that the secant becomes the tangent when the two points of intersection are one, it was felt that he had not overcome the difficulties inherent in Zeno's paradoxes. After all, does a variable quantity reach its limit? Or does it never reach it?

We have already referred to Bishop Berkeley's criticism of Newton's fluxions. George Berkeley, Dean of Derry in 1724, Bishop of Cloyne in S. Ireland in 1734——and from 1729 to 1731 a resident of Newport, R.I.—is primarily known for his extreme idealism (*esse est percipi*). He resented the support that Newtonian science gave to materialism, and he attacked the theory of fluxions in *The Analyst* (1734). He derided the infinitesimals as "ghosts of departed quantities"; if x receives an increment o, then the increment of x^n, divided by o, is

$$nx^{n-1} + \frac{n(n-1)}{1.2} x^{n-2} o + \cdots .$$

This is obtained by supposing o different from zero. The fluxion of x^n, nx^{n-1}, however, is obtained by taking o as equal to zero, when the hypothesis is suddenly shifted, since o was supposed to be different from zero. This was the "manifest sophism" which Berkeley discovered in the calculus, and he believed that its correct results were obtained by a compensation of errors. Fluxions were logically unaccountable. "But he who can digest a second or third Fluxion, a second or third Difference," Berkeley exclaimed to the "infidel mathematician" whom he addressed (Halley), "need not, methinks, be squeamish about any Point in Divinity." It has not been the only case in which a critical difficulty in a science has been used to strengthen an idealist philosophy.

John Landen, a self-taught British mathematician whose name is preserved in the theory of elliptic integrals, tried to overcome the basic difficulties in the calculus in his own way. In the *Residual Analysis* (1764) he met Berkeley's criticism by avoiding infinitesimals altogether; the derivative of x^3, for instance, was found by changing x into x_1, after which

$$\frac{x_1^3 - x^3}{x_1 - x} = x_1^2 + xx_1 + x^2$$

becomes $3x^2$ when $x = x_1$. Since this procedure involves infinite series when the functions are more complicated, Landen's method has some affinity to the later "algebraic" method of Lagrange.

6

Although Euler was incontestably the leading mathematician of this period, France continued to produce work of great originality. Here, more than in any other country, mathematics was conceived as the science which was to bring Newton's theory to greater perfection. The theory of universal gravitation had great attraction for the philosophers of the Enlightenment, who used it as a weapon in their struggle against the remnants of feudalism. The Catholic Church had placed Descartes on the Index in 1664, but by 1700 his theories had become fashionable even in conservative circles. The question of Newtonianism versus Cartesianism became for a while a topic of the greatest interest not only in learned circles but also in the salons. Voltaire's *Lettres sur les Anglais* (1734) did much to introduce Newton to the French reading public; Voltaire's friend Mme. du Châtelet even translated the *Principia* into French (1759). A particular point of contention between the two schools was the figure of the earth. In the cosmogony favored by the Cartesians the earth elongated at the poles; Newton's theory required that it be flattened. The Cartesian astronomers Cassini (Jean Dominique, the father, and Jacques, the son; the father known in geometry because of "Cassini's oval," 1680) had measured an arc of the meridian in France between 1700 and 1720, which vindicated the Cartesian conclusion. A controversy arose in which many mathematicians participated. An expedition was sent in 1735 to Peru,[11] followed in 1736–37 by one under the direction of Pierre de Maupertuis to the Torne in Sweden among the Lapps in order to measure a degree of longitude. The result of both expeditions was a triumph for Newton's theory, as well as for Maupertuis himself. The now famous *grand aplatisseur* ("great flattener") became president of the Berlin Academy and basked for many years in the sun of his fame at the court of Frederick the Great. This lasted until 1750, when he entered into a spirited controversy with the Swiss mathematician Samuel König concerning the principle of least action in mechanics, perhaps already indicated by Leibniz. Maupertuis was looking, as Fermat had done before and Einstein did after him, for some general principle by which the laws of the universe could be unified. Maupertuis's formulation was not clear, but he defined as his "action" the quantity mvs (m = mass, v = velocity, s = distance); he combined with it a proof of the existence of God. The controversy was brought to a climax when Voltaire lampooned the unhappy president in the *Diatribe du docteur Akakia, médecin du pape* (1752). Neither the king's support nor Euler's defense could bring succor to Maupertuis's sunken spirits, and the deflated mathematician died not long afterwards in Basel in the home of the Bernoullis.

Euler restated the principle of least action in the form that $\int mvds$ must be a minimum; moreover, he did not indulge in Maupertuis's metaphysics. This placed the principle on a sound basis, where it was used by Lagrange[12] and later by Hamilton. The use of the "Hamiltonian" in modern mathematical physics illustrates the fundamental character of Euler's contribution to the Maupertuis-König controversy.[13]

---

[11]This was the Spanish viceroyalty of Peru, much larger than the present Peru. Headquarters of the expedition were at Quito, now in Ecuador.

[12]See E. Mach, *The Science of Mechanics* (Chicago, 1893), p. 364; R. Dugas, *Histoire de la mécanique* (Neuchâtel, 1950).

[13]Details of the controversy are given in Euler, *Opera omnia*, 2nd ser., Vol. 5 (1957), introduction by J. O. Fleckenstein. The best authority on the whole subject of eighteenth-century mechanics is C. Truesdell; see "Literature," below.

Among the mathematicians who went with Maupertuis to Lapland was Alexis-Claude Clairaut. Clairaut, at eighteen years of age, had published the *Recherches sur les courbes à double courbure* (1731), a first attempt to deal with the analytical and differential geometry of space curves. Upon his return from Lapland, Clairaut published his *Théorie de la figure de la terre* (1743), a standard work on the equilibrium of fluids and the attraction of ellipsoids of revolution. Laplace could only improve on it in minor details. Among its many results is the condition that a differential $Mdx + Ndy$ be exact. This book was followed by the *Théorie de la lune* (1752), which added material to Euler's theory of the moon's motion and the problem of three bodies in general. Clairaut also contributed to the theory of line integrals and of differential equations; one of the types which he considered is known as "Clairaut's equation." It offered one of the first known examples of a singular solution. Clairaut also stated the theorem that for $z = f(x,y)$ the values $\partial^2 z / \partial x\ \partial y$ and $\partial^2 z / \partial y\ \partial x$ are equal. But the first to state it was Nikolaus Bernoulli (1721) and we also find it in Euler's *Introductio*.

7

The intellectual opposition to the *ancien régime* centered after 1750 around the famous *Encylopédie* (28 vols., 1751–72). The editor was Denis Diderot, under whose leadership the Encyclopedia presented a detailed philosophy of the Enlightenment. Diderot's knowledge of mathematics was not inconsiderable,[14] but the leading mathematician of the Encyclopedists was Jean Le Rond d'Alembert, the natural son of an aristocratic lady, left as a foundling near the church of St. Jean Le Rond in Paris. His early brilliance facilitated his career; in 1754 he became *secrétaire perpétuel* of the French Academy and as such the most influential man of science in France. In 1743 appeared his *Traité de dynamique*, which contains the method of reducing the dynamics of solid bodies to statics, known as "d'Alembert's principle." He continued to write on many applied subjects, especially on hydrodynamics, aerodynamics, and the three-body problem. In 1747 appeared his theory of vibrating strings which made him, together with Daniel Bernoulli, the founder of the theory of partial differential equations. Where d'Alembert and Euler solved the equation $z_{tt} = k^2 z_{xx}$ by means of the expression $z = f(x + kt) + \phi(x - kt)$, Bernoulli solved it by means of a trigonometric series. There remained

---

[14]There exists a widely quoted story about Diderot and Euler according to which Euler, in a public debate in St. Petersburg, succeeded in embarrassing the freethinking Diderot by claiming to possess an algebraic demonstration of the existence of God: "Sir, $(a + b^n)/n = x$; hence God exists, answer please!" This is a good example of a bad historical anecdote, since the value of an anecdote about an historical person lies in its faculty to illustrate certain aspects of his character; this particular anecdote serves to obscure the character both of Diderot and of Euler. Diderot knew his mathematics and had written on involutes and probability, and no reason exists to think that the thoughtful Euler would have behaved in the asinine way indicated. The story seems to have been made up by the English mathematician De Morgan (1806–1871). See L. G. Krakeur and R. L. Krueger, *Isis*, Vol. 31 (1940), pp. 431–32; also Vol. 33 (1941), pp. 219–31. It is true that there was in the eighteenth century occasional talk about the possibility of an algebraic demonstration of the existence of God; Maupertuis indulged in one, see Voltaire's *Diatribe, Oeuvres*, Vol. 41 (1821 ed.), pp. 19, 30. See also B. Brown, *Amer. Math. Monthly*, Vol. 49 (1944), and R. J. Gillings, *Amer. Math. Monthly*, Vol. 61 (1954), pp. 77–80.

grave doubts concerning the nature of this solution; d'Alembert believed that the initial form of the string could only be given by a single analytical expression, while Euler thought that "any" continuous curve would do. Bernoulli believed, contrary to Euler, that his series solution was perfectly general. The full explanation of the problem had to wait until 1824, when Fourier removed the doubts concerning the validity of a trigonometric series as the representation of "any" function.[15]

D'Alembert was a facile writer on many subjects, including even fundamental questions in mathematics. We mentioned his introduction of the limit conception. The "fundamental" theorem in algebra is sometimes called "d'Alembert's theorem" because of his attempt at proof (1746); and d'Alembert's "paradox" in the theory of probability shows that he also thought about the foundations of this theory—if not always very successfully.

The theory of probabilities made rapid advances in this period, mainly by further elaboration of the ideas of Fermat, Pascal, and Huygens. The *Ars conjectandi* was followed by several other texts, among them *The Doctrine of Chances* (1716), written by Abraham de Moivre, a French Huguenot who settled in London after the revocation of the Edict of Nantes (1685) and earned a living by private tutoring. De Moivre's name is attached to a theorem in trigonometry, which in its present form $(\cos \varphi + i \sin \varphi)^n = \cos n\varphi + i \sin n\varphi$ appears first in Euler's *Introductio*. In a paper published in 1733 he derived the normal probability function as an approximation to the binomial law and a formula equivalent to that of Stirling. James Stirling, a Scotch mathematician of the Newtonian school, published his series in 1730.

The many lotteries and insurance companies that were organized in this period interested many mathematicians, including Euler, in the theory of probabilities. It led to attempts to apply the doctrine of chances to new fields. Georges-Louis Leclerc, Comte de Buffon, director of the Paris Jardin du Roi, who is noted as the author of a natural history in thirty-six delightful volumes and the famous discourse on style (1753: *le style c'est l'homme même*), introduced in 1777 the first example of a geometrical probability. This was the so-called needle problem, which has appealed to the imagination of many people because it allows the "experimental" determination of π by throwing a needle on a plane covered with parallel and equidistant lines and counting the number of times the needle hits a line.

To this period belongs also the attempt to apply the theory of probability to man's judgment; for instance, by computing the chance that a tribunal can arrive at a true verdict if to each of the different witnesses and jurymen a number can be given expressing the chance that he will speak or understand the truth. This curious *probabilité des jugements*, with its distinct flavor of Enlightenment philosophy, was prominent in the work of the Marquis de Condorcet; it reappeared in Laplace and even in Poisson (1837).

---

[15]For the development of the function concept, see A. P. Youschkevitch (Juškevič), "The Concept of Function up to the Middle of the 19th Century," *AHES*, Vol. 16 (1976), pp. 37–85. Further: A. F. Monna, *AHES*, Vol. 9 (1972), pp. 57–84.

8

De Moivre, Stirling, and Landen were good representatives of British eighteenth-century mathematicians. We must report on a few more, although none of them reached the heights of their continental colleagues. The tradition of the venerated Newton rested heavily upon English science, and the clumsiness of his notation as compared to that of Leibniz made progress difficult. There were deep-lying social reasons why English mathematicians refused to be emancipated from Newtonian fluxional methods. England was in constant commercial wars with France and developed a feeling of intellectual superiority which was fostered not only by its victories in war and trade but also by the admiration in which the continental philosophers held its political system. England became the victim of its own supposed excellence. An analogy exists between the mathematics of eighteenth-century England and of late Alexandrian antiquity. In both cases progress was technically impeded by an inadequate notation, but reasons for the self-satisfaction of the mathematicians were of a deeper-lying social nature.

The leading British mathematician of this period was the Scotchman Colin Maclaurin, professor at the University of Edinburgh, a disciple of Newton, with whom he had been personally acquainted. His study and extension of fluxional methods, of curves of second and higher order, and of the attraction of ellipsoids run parallel with contemporary efforts of Clairaut and Euler. Several of Maclaurin's theorems occupy a place in our theory of plane curves and our projective geometry. In his *Geometria organica* (1720) we find the observation known as "Cramer's paradox" that a curve of the nth order is not always determined by $1/2n(n + 3)$ points, so that nine points may not uniquely determine a cubic while ten would be too many. Here we also find kinematical methods to describe plane curves of different degrees. Maclaurin's *Treatise of Fluxions* (2 vols., 1742)—written to defend Newton against Berkeley—is difficult to read because of the antiquated geometrical language, which is in sharp contrast with the ease of Euler's writing. Maclaurin used this method in an attempt to obtain Archimedean rigor. The book contains Maclaurin's investigations on the attraction of ellipsoids of revolution and his theorem that two such ellipsoids, if they are confocal, attract a particle on the axis or in the equator with forces proportional to their volumes. In this *Treatise* Maclaurin also deals with the famous "series of Maclaurin."

This series, however, was no new discovery, since it had appeared in the *Methodus incrementorum* (1715) written by Brook Taylor, for a while secretary of the Royal Society. Maclaurin fully acknowledged his debt to Taylor. The series of Taylor is now always given in Lagrange's notation:

$$f(x + h) = f(x) + h f'(x) + \frac{h^2}{2!} f''(x) + \cdots;$$

Taylor explicitly mentions the series for $x = 0$, which many college texts still insist on naming "Maclaurin's series." Taylor's derivation did not include convergence considerations, but Maclaurin made a beginning with such considerations—he even had the so-called integral test for infinite series. The full impor-

LEONHARD EULER (1707–1783)
(From the portrait by A. Lorgna.)

JEAN LE ROND D'ALEMBERT (1717–1783)

JOSEPH-LOUIS LAGRANGE
(1736–1813)

PIERRE-SIMON LAPLACE
(1749–1827)
(From an engraving after
a painting by Naigeon.)

JEAN-ETIENNE MONTUCLA (1725–1799)
(After an engraving after a miniature by P. Viel.)

tance of Taylor's series was not recognized until Euler applied it in his differential calculus (1755). Lagrange supplied it with the remainder and used it as the foundation of his theory of functions. Taylor himself used his series for the integration of some differential equations. He also began the study of the vibrating string, which was taken up later by d'Alembert and others (see Sec. 7, above).

9

Joseph-Louis Lagrange was born in Turin of Italian-French ancestry. At nineteen years of age he became professor of mathematics in the artillery school of Turin (1755). In 1766, when Euler left Berlin for St. Petersburg, Frederick the Great invited Lagrange to come to Berlin, accompanying his invitation, it is said, with the modest message that "it is necessary that the greatest geometer of Europe should live near the greatest of kings." Lagrange stayed at Berlin until the death of Frederick (1786), after which he went to Paris. During the Revolution he assisted in reforming weights and measures; later he became professor, first at the Ecole Normale (1795), then at the Ecole Polytechnique (1797).

To Lagrange's earliest works belong his contributions to the calculus of variations. Euler's memoir on this subject had appeared in 1755. Lagrange observed that Euler's method had "not all the simplicity which is desirable in a subject of pure analysis." The result was Lagrange's purely analytical calculus of variations (1760–61), which is not only full of original discoveries but also has the historical material well arranged and assimilated—something quite typical of all Lagrange's work. Lagrange immediately applied his theory to problems of dynamics, in which he made full use of Euler's formulation of the principle of least action, the result of the lamentable *Akakia* episode (see p. 127). Many of the essential ideas of the *Mécanique analytique* thus date back to Lagrange's Turin days. He also contributed to one of the standard problems of his day, the theory of the moon, important not only for its own sake, but for the solution of the problem of finding longitude. Lagrange gave the first particular solutions of the three-body problem. The theorem of Lagrange states that it is possible to start three finite bodies in such a manner that their orbits are similar ellipses all described in the same time (1772). In 1767 appeared his memoir, *Sur la résolution des équations numériques*, in which he presented methods of separating the real roots of an algebraic equation and of approximating them by means of continued fractions. This was followed in 1770 by the *Réflexions sur la résolution algébrique des équations*, which dealt with the fundamental question of why the methods useful to solve equations of degree $n \leq 4$ are not successful for $n > 4$. This led Lagrange to rational functions of the roots and their behavior under the permutations of the roots, the procedure which not only stimulated Ruffini and Abel in their work on the case $n > 4$, but also led Galois to his theory of groups. Lagrange also made progress in the theory of numbers when he investigated quadratic residues and proved, among many other theorems, that every integer is the sum of four or less than four squares.

Lagrange devoted the second part of his life to the composition of his great works: *Mécanique analytique* (1788), *Théorie des fonctions analytiques* (1797), and its sequel, *Leçons sur le calcul des fonctions* (1801). The two books on functions were an attempt to give a solid foundation to the calculus by reducing it to algebra. Lagrange rejected the theory of limits as indicated by Newton and formulated by d'Alembert. He could not well understand what happened when $\Delta y/\Delta x$ reaches its limit. In the words of Lazare Carnot, the *organisateur de la victoire* in the French Revolution, who also worried about Newton's method of infinitesimals:

> That method has the great inconvenience of considering quantities in the state in which they cease, so to speak, to be quantities, for though we can always well conceive the ratio of two quantities, as long as they remain finite, that ratio offers to the mind no clear and precise idea, as soon as its terms become, the one and the other, nothing at the same time.[16]

Lagrange's method was different from that of his predecessors. He started with Taylor's series, which he derived with their remainder, showing in a rather naïve way that "any" function $f(x)$ could be developed in such a series with the aid of a purely algebraic process. Then the derivatives $f'(x)$, $f''(x)$, etc., were defined as the coefficients $h, h^2, \ldots$ in the Taylor expansion of $f(x + h)$ in terms of h. [The notation $f'(x), f''(x)$ is due to Lagrange.]

Though this "algebraic" method of founding the calculus turned out to be unsatisfactory, and though Lagrange gave insufficient attention to the convergence of the series, the abstract treatment of a function was a considerable step ahead. Here appeared a first "theory of functions of a real variable" with applications to a large variety of problems in algebra and geometry.

Lagrange's *Mécanique analytique* is perhaps his most valuable work and still amply repays careful study. In this book, which appeared a hundred years after Newton's *Principia*, the full power of the newly developed analysis was applied to the mechanics of points and of rigid bodies. The results of Euler, d'Alembert, and the other mathematicians of the eighteenth century were assimilated and further developed from a consistent point of view. Full use of Lagrange's own calculus of variations made the unification of the varied principles of statistics and dynamics possible—in statistics by the use of the principle of virtual velocities, in dynamics by the use of d'Alembert's principle. This led naturally to generalized coordinates and to the equation of motion in their "Lagrangian" form:

$$\frac{d}{dt}\frac{\partial T}{\partial q_i} - \frac{\partial T}{\partial q_i} = F_i.$$

Newton's geometrical approach was now fully discarded; Lagrange's book was a triumph of pure analysis. The author went so far as to stress in the preface: *On ne trouvera point de figures dans cet ouvrage, seulement des opérations algébriques.*[17] It characterized Lagrange as the first true analyst.

---

[16]L. Carnot, *Réflexions sur la métaphysique du calcul infinitésimal*, 5th ed. (Paris, 1881), p. 147; see also F. Cajori, *Amer. Math. Monthly*, Vol. 22 (1915), p. 148.

[17]"No figures will be found in this work, only algebraic operations." The word "algebraic" instead of "analytic" is characteristic.

10

With Pierre-Simon Laplace we reach the last of the leading eighteenth-century mathematicians. The son of a Normandy farmer, he attended classes at Beaumont and Caen, and through the aid of d'Alembert became professor of mathematics at the military school of Paris. He had several other teaching and administrative positions and took part during the Revolution in the organization of the Ecole Normale as well as the Ecole Polytechnique. Napoleon bestowed many honors upon him, but so did Louis XVIII. In contrast to Monge and Carnot, Laplace easily shifted his political allegiances and besides was somewhat of a snob; but this easy conscience enabled him to continue his purely mathematical activity despite all political changes in France.

The two great works of Laplace which unify not only his own investigations, but all previous work in their respective subjects, are *Théorie analytique des probabilités* (1812) and *Mécanique céleste* (5 vols., 1799–1825). Both monumental works were prefaced by extensive expositions in nontechnical terms, the *Essai philosophique sur les probabilités* (1814) and *Exposition du système du monde* (1796). This *Exposition* contains the nebular hypothesis, independently proposed by Kant in 1755 (and even before Kant by Swedenborg in 1734). The *Mécanique céleste* was the culmination of the work of Newton, Clairaut, d'Alembert, Euler, Lagrange, and Laplace on the figure of the earth, the theory of the moon, the three-body problem, and the perturbations of the planets, leading up to the momentous problem of the stability of the solar system. The name "Laplace's equation," given to

$$\frac{\partial^2 V}{\partial x^2} + \frac{\partial^2 V}{\partial y^2} + \frac{\partial^2 V}{\partial z^2} = 0,$$

reminds us that potential theory is part of the *Mécanique céleste*. (The equation itself had already been found by Euler in 1752 when he derived some of the principal equations of hydrodynamics.) Around this five-volume opus cluster many anecdotes. Well-known is Laplace's supposed answer to Napoleon, who tried to tease him by the remark that God was not mentioned in his book: *"Sire, je n'avais pas besoin de cette hypothèse."*[18] And Nathaniel Bowditch of Boston, who translated four volumes of Laplace's work into English, remarked: "I never came across one of Laplace's 'Thus it plainly appears' without feeling sure that I had hours of hard work before me to fill up the chasm and find out and show how it plainly appears." Hamilton's mathematical career began by finding a mistake in Laplace's *Mécanique céleste*. Green, reading Laplace, conceived the idea of a mathematical theory of electricity.

The *Essai philosophique sur les probabilités* is a very readable introduction to the theory of probabilities; it contains Laplace's "negative" definition of probabilities by postulating "equally likely events":

> The theory of chance consists in the reduction of all events of the same kind to a certain number of equally likely cases, that are cases such that we are equally

---

[18]"Sire, I had no need of that hypothesis."

undecided about their existence, and in determining the number of cases which are favorable to the event of which we seek the probability.

Questions concerning probability appear, according to Laplace, because we are partly ignorant and partly knowing. This led Laplace to his famous statement which summarizes the eighteenth-century interpretation of mechanical materialism:

> An intelligence which, for a given instant, knew all the forces by which nature is animated and the respective position of the beings which compose it, and which besides was large enough to submit these data to analysis, would embrace in the same formula the motions of the largest bodies of the universe and those of the lightest atom: nothing would be uncertain to it, and the future as well as the past would be present to its eyes. Human mind offers a feeble sketch of this intelligence in the perfection which it has been able to give to Astronomy.

The standard text itself is so full of material that many later discoveries in the theory of probabilities can already be found in Laplace.[19] The stately volume contains an extensive discussion of games of chance and of geometrical probabilities, of Bernoulli's theorem and of its relation to the normal integral, and of the theory of least squares invented by Legendre. The leading idea is the use of the *fonctions génératrices*, of which Laplace shows the power for the solution of difference equations. It is here that the "Laplace transform" is introduced, which later became the key to the Heaviside operational calculus. Laplace also rescued from oblivion and reformulated a theory sketched by Thomas Bayes, an obscure English clergyman, which was posthumously published in 1763–64. This theory became known as the theory of inverse probabilities.

11

It is a curious fact that toward the end of the century some of the leading mathematicians expressed the feeling that the field of mathematics was somehow exhausted. The laborious efforts of Euler, Lagrange, d'Alembert, and others had already led to the most important theorems; the great standard texts had placed them, or would soon place them, in their proper setting; the few mathematicians of the next generation would only find minor problems to solve. *Ne vous semble-t-il pas que la haute géométrie va un peu à décadence?* wrote Lagrange to d'Alembert in 1772. *Elle n'a d'autre soutien que vous et M. Euler.*[20] Lagrange even discontinued working in mathematics for a while. D'Alembert had little hope to give. Arago, in his *Éloge de Laplace* (1842), later expressed a sentiment which may help us to understand this feeling:

> Five geometers—Clairaut, Euler, d'Alembert, Lagrange and Laplace—shared between them the world of which Newton had revealed the existence. They

---

[19]E. C. Molina, "The Theory of Probability: Some Comments on Laplace's 'Théorie analytique,'" *Bull. Am. Math. Soc.*, Vol. 36 (1930), pp. 369–92.
[20]"Does it not seem to you that sublime geometry is tending to become a little decadent? It has no other support than you and Mr. Euler." "Geometry" in eighteenth-century French is used for mathematics in general.

explored it in all directions, penetrated into regions believed inaccessible, pointed out countless phenomena in those regions which observation had not yet detected, and finally—and herein lies their imperishable glory—they brought within the domain of a single principle, a unique law, all that is most subtle and mysterious in the motions of the celestial bodies. Geometry also had the boldness to dispose of the future; when the centuries unroll themselves they will scrupulously ratify the decisions of science.[21]

Arago's oratory points to the main source of this *fin de siècle* pessimism, which consisted of the tendency to identify the progress of mathematics too much with that of mechanics and astronomy. From the time of ancient Babylon until that of Euler and Laplace, astronomy had guided and inspired the most sublime discoveries in mathematics; now this development seemed to have reached its climax. However, a new generation, inspired by the new perspectives opened by the French Revolution and the flowering of the natural sciences, was to show how unfounded this pessimism was. This great new impulse came only in part from France; it also came, as often in the history of civilization, from the periphery of the political and economic centers, in this case from Gauss in Göttingen.

12

To the France of the Age of Enlightenment belongs the first comprehensive history of mathematics, a very readable narrative, not just a catalog of names and titles like those of the past. It is the *Histoire des mathématiques* by Jean-Etienne Montucla, a lawyer moving in the circle of the Encyclopedists, holding after 1761 a series of important government jobs. Published first in 1758 in two volumes, and later, with the aid of the astronomer J.-J. de Lalande, in four (1799–1802), it contains both "pure" and "mixed" (applied) mathematics, even music. The book, although it stimulated further research, had no successor until that of Moritz Cantor a hundred years later, and then without the applied part. Montucla's work was followed by the two-volume *Essai sur l'histoire générale des mathématiques* by the Abbé Charles Bossuet (1802), editor of Pascal's work (1779), which was widely read, also in translations.

Literature

The works of Lagrange and Laplace are available in modern editions; those of Euler are nearing completion in a monumental edition, many volumes containing important introductions. Volumes of Euler's correspondence have also been published. The edition of the works of all the mathematical Bernoullis has begun; among those published are the *Ars conjectandi* in *Die Werke von Jakob Bernoulli*, Band 3 (Basel, 1975) with commentary. Before this we only had an edition of Jakob's work (1744, 2 vols.) and of Johann's (1742, 4 vols.). The monumental edition of Euler's work, *Leonhardi Euleri Opera Omnia*, started in 1911; it is

---

[21] F. Arago, *Oeuvres complètes* (Paris and Leipzig, 1855), Vol. 3, p. 464.

published in three series; the first one, the *Opera Mathematica* in 29 volumes, has been completed. See A. P. Youschkevitch (Juškevič), *DSB*, Vol. 4 (1971), pp. 467–84.

Truesdell, C. "Leonhard Euler, Supreme Geometer (1707–1783)." *Studies in Eighteenth Century Culture*, Vol. 2 (1972), pp. 51–95.
Juškevič, A. P., and Winter, E. *Leonhard Euler und Christian Goldbach. Briefwechsel.* 1729–1764; Berlin, 1965.
See on Laplace the extensive article by C. C. Gillespie in *DSB*, Vol. 15, Suppl. I (1978), pp. 273–403.
[Mills, S., ed.] *The Collected Letters of Colin MacLaurin.* Nantwick, Cheshire, 1982 (Birkhäuser, Boston, distributor).
Lambert, J. H. *Opera mathematica.* 2 vols., Berlin, 1946. (Vol. 1, pp. ix–xxxi, contains a preface by A. Speiser.)
Cajori, F. *A History of the Conception of Limits and Fluxions in Great Britain from Newton to Woodhouse.* Chicago, 1931.
Jourdain, P. E. B. *The Principle of Least Action.* Chicago, 1913.
DuPasquier, L. G. *Léonard Euler et ses amis.* Paris, 1927.
Spiess, O. *Leonhard Euler.* Frauenfeld and Leipzig, 1929.
Leonard Euler, *Collection of Articles in Honor of the 250th Anniversary of His Birth.* Moscow, 1958 (in Russian). [See also articles in *Istor.-mat. Issled.*, Vol. 7 (1954), pp. 451–640 (in Russian). Also, *A Collection of Articles Published on the Occasion of the 150th Anniversary of Euler's Death*, Moscow and Leningrad, 1935 (in Russian).]
Euler, L. *Vollständige Anleitung zur Algebra*, J. E. Hofmann, ed. Reclam, Stuttgart, 1959 (with historical introduction).
Fleckenstein, J. O. "Johann und Jakob Bernoulli." In *Elemente der Mathematik*, Suppl. 7, Basel, 1949.
Hofmann, J. E. "Über Jakob Bernoulli's Beiträge zur Infinitesimalmathematik." *Enseignement mathématique*, 2nd ser., 5 (1956), pp. 61–171.
Andoyer, H. *L'oeuvre scientifique de Laplace.* Paris, 1922.
Loria, G. "Nel secondo centenario della nàscita di G. L. Lagrange." *Isis*, Vol. 28 (1938), pp. 366–75. (Contains full bibliography.)
Auchter, H. *Brook Taylor der Mathematiker und Philosoph.* Marburg, 1937. (Shows from data in Leibniz's manuscripts that Leibniz had the Taylor series by 1694. James Gregory already possessed it in 1671.)
Green, H. G., and Winter, H. J. J. "John Landen, F.R.S. (1719–90), Mathematician." *Isis*, Vol. 35 (1944), pp. 6–10.
[Bayes, T.] Facsimile of Two Papers, with commentaries by E. C. Molina and W. E. Deming. Washington, D.C., 1940.
Pearson, K. "Laplace." *Biometrica*, Vol. 21 (1929), pp. 202–16.
Stäckel, P. "Zur Geschichte der Funktionentheorie im achtzehnten Jahrhundert." *Bibliotheca mathematica*, Vol. 3, No. 2 (1901), pp. 111–21.
Nielsen, N. *Géomètres français du dix-huitième siècle.* Copenhagen and Paris, 1935.
Les oeuvres de Nicolas Struyck (1687–1769) qui se rapportent au calcul des chances, J. A. Vollgraf, ed. Amsterdam, 1912.
Sarton, G. "Montucla." *Osiris*, Vol. 1 (1936), pp. 519–67.
Maystrov, L. E. "Lomonossov, Father of Russian Mathematics." *The Soviet Review*, Vol. 3, No. 3 (1962), pp. 3–18. [Trans. from *Voprosy Filosofii*, Vol. 5 (1961).]
Scott, J. F. "Mathematics Through the Eighteenth Century." *Philos Mag.*, Commemoration Number (1948), pp. 67–90. (The stress is on England.)

Truesdell, C. *The Rational Mechanics of Flexible or Elastic Bodies, 1630–1780.* In Euler's *Opera Omnia,* 2nd ser., Vol. 11, Zurich, 1960. [Other important introductions to Euler's *Opera Omnia* are those by G. Faber on infinite series (1st ser., Vol. 16, 1935); G. Faber and A. Krazer on integrals (1st ser., Vol. 19, 1932); C. Carathéodory on variational calculus (1st ser., Vol. 24, 1952); A. Speiser on geometry (1st ser., Vols. 26–29, 1953–56); and J. O. Fleckenstein on mechanics (2nd ser., Vol. 5, 1957).]

Schneider, I. "Der Mathematiker Abraham de Moivre (1667–1754)." *AHES,* Vol. 5 (1968), 177–317.

Sheynin, O. B. "R. J. Boscovitch's Work on Probability." *AHES,* Vol. 9 (1973), pp. 306–24. (Rudjev Josip Boškovič [1711–87], Croatian-born Jesuit, known as a "polymath.")

Brunet, P. "La vie et l'oeuvre de Clairaut." *Revue d'histoire des sciences,* Vol. 4 (1951), pp. 13–40, 109–53; also Paris (1952).

Gillespie, C. C. *Lazare Carnot, savant.* Princeton, N.J., 1970.

CHAPTER VIII

The Nineteenth Century

1

The French Revolution and the Napoleonic period created extremely favorable conditions for the further growth of mathematics. The way was open for the Industrial Revolution on the continent of Europe. It stimulated the cultivation of the physical sciences; it created new social classes with a new outlook on life, interested in science and in technical education. Democratic ideas invaded academic life; criticism rose against antiquated forms of thinking; schools and universities had to be reformed and rejuvenated.

The new and turbulent mathematical productivity was not primarily due to the technical problems raised by the new industries. England, the heart of the Industrial Revolution, remained mathematically almost sterile for several decades. Mathematics progressed most healthily in France and somewhat later in Germany, countries in which the ideological break with the past was most sharply felt and where sweeping changes were made, or had to be made, to prepare the ground for the new capitalist economic and political structure. The new mathematical research gradually emancipated itself from the ancient tendency to see in mechanics and astronomy the final goal of the exact sciences. The pursuit of science as a whole became even more detached from the demands of economic life or of warfare. The specialist developed, interested in science for its own sake. The connection with practice was never entirely broken, but it often became obscured. A sharper division than in previous times between practitioners of "pure" and "applied" mathematics accompanied the growth of specialization.[1]

---

[1]The difference in approach found its classical expression in the remark by Jacobi on the opinions of Fourier, who still represented the utilitarian approach of the eighteenth century: "*Il est vrai que Monsieur Fourier avait l'opinion que le but principal des mathématiques était l'utilité publique et l'explication des phénomènes naturels; mais un philosophe comme lui aurait dû savoir que le but unique de la science, c'est l'honneur de l'esprit humain, et que sous ce titre une question de nombre vaut autant qu'une question du système du monde.*" ("It is true that Mr. Fourier believed that the main end of mathematics was public usefulness and the explanation of the phenomena in nature, but such a philosopher as he was should have known that the sole end of science is the honor of the human mind, and that from this point of view a question concerning number is as important as a question concerning the system of the world.") In a letter to Legendre in 1830 (*Werke*, I, p. 454), Gauss represented a synthesis of both opinions; he freely applied mathematics to astronomy, physics, and geodesy, but at the same time considered mathematics the "queen of the sciences," and arithmetic the "queen of mathematics."

The mathematicians of the nineteenth century were no longer at royal courts or in the salons of the aristocracy. Their chief occupation no longer consisted in membership in a learned academy; they were usually employed by universities or technical schools and were teachers as well as investigators. The Bernoullis, Lagrange, and Laplace had done occasional teaching. Now the teaching responsibility increased; mathematics professors became educators and examiners of youth. The internationalism of previous centuries tended to be undermined by the growing relationship between the scientists of each nation, though international exchange of opinion did remain. Scientific Latin was gradually replaced by the national languages. Mathematicians began to work in specialized fields; and while Leibniz, Euler, and d'Alembert can be described as "mathematicians" (as *géomètres* in the eighteenth-century meaning of the word), we think of Cauchy as an analyst, of Cayley as an algebraist, of Steiner as a geometer (even a "pure" geometer), and of Cantor as a pioneer in point set theory (theory of aggregates). The time was ripe for "mathematical physicists" followed by men learned in "mathematical statistics" or "mathematical logic." Specialization was only broken on the highest level of genius; and it was from the works of a Gauss, a Riemann, a Klein, a Poincaré that nineteenth-century mathematics received its most powerful impetus.

2

On the dividing line between eighteenth- and nineteenth-century mathematics towers the majestic figure of Carl Friedrich Gauss. He was born in the German city of Brunswick, the son of a day laborer. The Duke of Brunswick gracefully recognized an infant prodigy in young Gauss and took charge of his education. The young genius studied from 1795 to 1798 at Göttingen and in 1799 obtained his doctor's degree at Helmstedt. From 1807 until his death in 1855 he worked quietly and undisturbed as the director of the astronomical observatory and professor at his alma mater. His comparative isolation, his grasp of "applied" as well as "pure" mathematics, his preoccupation with astronomy and his frequent use of Latin have the touch of the eighteenth century, but his work breathes the spirit of a new period. He stood, with his contemporaries Kant, Goethe, Beethoven, and Hegel, on the sidelines of a great political struggle raging in other countries, but expressed in his own field the new ideas of his age in a most powerful way.

Gauss's diaries show that in his seventeenth year he had already begun to make startling discoveries. In 1795, for instance, he discovered independently of Euler the law of quadratic reciprocity in number theory. Some of his early discoveries were published in his Helmstedt dissertation of 1799 and in the impressive *Disquisitiones arithmeticae* of 1801. The dissertation gave the first rigorous proof of the so-called "fundamental theorem of algebra," which states that every algebraic equation with real coefficients has at least one root and hence has n roots. The theorem may go back to Albert Girard, the editor of Stevin's works (*Invention nouvelle en algèbre*, 1629); d'Alembert had tried to give a proof in 1746. Gauss loved this theorem and later gave two more demonstrations, returning in the fourth (1849) to his first proof. The third demonstration (1816) used complex integrals and showed Gauss's early mastery of the theory of complex numbers.

The *Disquisitiones arithmeticae* collected all the masterful work in number theory of Gauss's predecessors and enriched it to such an extent that the beginning of modern number theory is sometimes dated from the publication of this book. Its core is the theory of quadratic congruences, forms, and residues; it culminates in the law of quadratic residues, that *theorema aureum* for which Gauss gave the first complete proof. Gauss was as fascinated by this theorem as by the fundamental theorem of algebra, and later published five more demonstrations; one more was found among his papers after his death. The *Disquisitiones* also contain Gauss's studies on the division of the circle, in other words, on the roots of the equation $x^n = 1$. They led to the remarkable theorem that the sides of the regular polygon of 17 sides (more generally, of n sides, $n = 2^p + 1$, $p = 2^k$, n prime, $k = 0, 1, 2, 3 \ldots$) can be constructed with compass and ruler alone, a striking extension of the Greek type of geometry.

Gauss's interest in astronomy was aroused when, on the first day of the new century (January 1, 1801), Piazzi in Palermo discovered the first planetoid, which was given the name of Ceres. Since only a few observations of the new planetoid could be made, the problem arose to compute the orbit of a planet from a smaller number of observations. Gauss solved the problem completely; it leads to an equation of degree eight. When in 1802, Pallas, another planetoid, was discovered, Gauss began to take interest in the secular perturbations of planets. This led to the *Theoria motus corporum coelestium* (1809),[2] to his paper on the attraction of general ellipsoids (1813), to his work on mechanical quadrature (1814), and to his study of secular perturbations (1818). To this period belongs also Gauss's paper on the hypergeometric series (1812), which allows a discussion of a large number of functions from a single point of view. It is the first systematic investigation into the convergence of a series.

After 1820 Gauss began to be actively interested in geodesy. Here he combined extensive applied work in triangulation with his theoretical research. One of the results was his exposition of the method of least squares (1821, 1823), already the subject of investigation by Legendre (1806) and Laplace. Perhaps the most important contribution of this period in Gauss's life was the surface theory of the *Disquisitiones generales circa superficies curvas* (1827), which approached its subject in a way strikingly different from that of Monge. Here again practical considerations, now in the field of higher geodesy, were intimately connected with subtle theoretical analysis. In this publication appeared the so-called intrinsic geometry of a surface, in which curvilinear coordinates are used to express the linear element ds in a quadratic differential form $ds^2 = E\,du^2 + F\,du\,dv + G\,dv^2$. There was also a climax, the *theorema egregium*, which states that the total curvature of the surface depends only on E, F, and G and their derivatives, and thus is a bending invariant. But Gauss did not neglect his first love, the "queen of mathematics," even in this period of concentrated activity on problems of geodesy, for in 1825 and 1831 appeared his work on biquadratic residues. It was a continuation of this theory of quadratic residues in the *Disquisitiones arithmeticae*,

---

[2]Translated as *Theory of the Motion of the Heavenly Bodies Moving about the Sun in Conic Sections* by Charles Henry Davis.

CARL FRIEDRICH GAUSS (1777–1855)
(After a painting by A. Jensen.)

ADRIEN-MARIE LEGENDRE (1752-1833)

but a continuation with the aid of a new method, the theory of complex numbers. The treatise of 1831 gave not only an algebra of complex numbers, but also an arithmetic. A new prime-number theory appeared, in which 3 remains prime but $5 = (1 + 2i)(1 - 2i)$ is no longer a prime. This new complex-number theory clarified many dark points in arithmetic so that the law of quadratic reciprocity became simpler than in real numbers. In this paper Gauss dispelled forever the mystery that still surrounded complex numbers by his representation of them by points in a plane.[3]

A statue in Göttingen represents Gauss and his younger colleague Wilhelm Weber in the process of inventing the electric telegraph. This happened in 1833–34 at a time when Gauss's attention began to turn to physics. In this period he did much experimental work on terrestrial magnetism. But he also found time for a theoretical contribution of the first importance—his *Allgemeine Lehrsätze*, on the theory of forces acting inversely proportional to the square of the distance (1839, 1840). This was the beginning of potential theory as a separate branch of mathematics (Green's paper in 1828 was practically unknown at that time), and it led to certain minimal principles concerning space integrals, in which we recognize "Dirichlet's principle." For Gauss the existence of a minimum was evident; this later became a much debated question which was finally solved by Hilbert.

Gauss remained active until his death in 1855. In his later years he concentrated more and more on applied mathematics. His publications, however, do not give an adequate picture of his full greatness. The appearance of his diaries and of some of his letters has shown that he kept some of his most penetrating thoughts to himself. We now know that Gauss, as early as 1800, had discovered elliptic functions and around 1816 was in possession of non-Euclidean geometry. He never published anything on these subjects; indeed, only in some letters to friends did he disclose his critical position toward attempts to prove Euclid's parallel axiom. Gauss seems to have been unwilling to venture publicly into any controversial subject. In letters he wrote about the wasps who would then fly around his ears and of the "shouts of the Boeotians" that would be heard if his secrets were not kept. For himself, Gauss doubted the validity of the accepted Kantian doctrine that space conception is Euclidean *a priori*; for him the real geometry of space was a physical fact to be discovered by experiment.

3

In his history of mathematics of the nineteenth century Felix Klein has invited comparison between Gauss and the twenty-five-years-older French mathematician, Adrien-Marie Legendre. It is perhaps not entirely fair to compare Gauss with any mathematician except the very greatest; but this particular comparison shows how Gauss's ideas were "in the air," since Legendre in his own independent way worked on most subjects that occupied Gauss. Legendre taught from 1775 to

[3]Cf. E. T. Bell, "Gauss and the Early Development of Algebraic Numbers," *Nat. Math. Mag.*, Vol. 18 (1944), pp. 188, 219. Euler and other mathematicians after 1760 had already thought in terms of such a representation of complex numbers. See A. Speiser, "Introduction to Euler," *Opera omnia*, 1st ser., Vol. 28 (1955), p. xxxvii.

1780 at the military school in Paris and later filled several different government positions, such as professor at the Ecole Normale, examiner at the Ecole Poly-technique, and geodetic surveyor.

Like Gauss he did fundamental work on number theory (*Essai sur les nombres*, 1798; *Théorie des nombres*, 1830), in which he gave a formulation of the law of quadratic reciprocity. He also did important work on geodesy and on theoretical astronomy, was as assiduous a computer of tables as Gauss, formulated in 1806 the method of least squares, and studied the attraction of ellipsoids—even those that are not surfaces of revolution. Here he introduced the "Legendre functions." He also shared Gauss's interest in elliptic and Eulerian integrals as well as in the foundations and methods of Euclidean geometry.

Although Gauss penetrated more deeply into the nature of all these different fields of mathematics, Legendre produced works of outstanding importance. His comprehensive textbooks were for a long time authoritative, especially his *Exercices du calcul intégral* (3 vols., 1811-19) and his *Traité des fonctions ellip-tiques et des intégrals euleriennes* (1827-32), which is still a standard work. In his *Éléments de géométrie* (1794) he broke with the Platonic ideals of Euclid and presented a textbook of elementary geometry based on the requirements of modern education. This book passed through many editions and was translated into several languages; it has had a lasting influence.

4

The beginnings of the new period in the history of French mathematics may be dated perhaps from the establishment of military schools and academies, which took place during the latter part of the eighteenth century. These schools, of which some were also founded outside of France (Turin, Woolwich), paid consid-erable attention to the teaching of mathematics as a part of the training of military engineers. Lagrange started his career at the Turin school of artillery; Legendre and Laplace taught at the military school in Paris, Monge at Mézières. Carnot was a captain of engineers. Napoleon's interest in mathematics dates back to his student days at the military academies of Brienne and Paris. During the invasion of France by the Royalist armies the need for more centralized instruction in military engineering became apparent. This led to the foundation of the Ecole Polytechnique of Paris (1794), a school which soon developed into a leading institution for the study of general engineering and eventually became the model for all engineering and military schools of the early nineteenth century, including West Point in the United States. Instruction in theoretical and applied mathemat-ics was an integral part of the curriculum. Emphasis was laid upon research as well as upon teaching. The best scientists of France were induced to lend their support to the school; many great French mathematicians were students, pro-fessors, or examiners at the Ecole Polytechnique.[4]

---

[4]Cf. C. G. J. Jacobi, *Werke*, Vol. 7, p. 355 (lecture held in 1835). The Polytechnique has been made the subject of many studies relating to its role in shaping the mathematics of the early nineteenth century, both in research and education. Cf. H. Wussing, *Pädagogik*, Vol. 13 (1958), pp. 646–62. For background, see M. P. Crosland, *The Society of Arcueil* (Cambridge, Mass., 1967). Laplace's residence at Arcueil, near Paris, was from 1806 to 1813 a center of social and scientific discourse.

The instruction at this institution, as well as at other technical schools, required a new type of textbook. The learned treatises for the initiated that were so typical of Euler's period had to be supplemented by college handbooks. Some of the best textbooks of the early nineteenth century were prepared for the instruction at the Ecole Polytechnique or similar institutions. Their influence can be traced in our present-day texts. A good example of such a handbook is the *Traité du calcul différentiel et du calcul intégral* (2 vols., 1797), written by Sylvestre-François Lacroix, from which whole generations have learned calculus. We have already mentioned Legendre's books. A further example is Monge's textbook of descriptive geometry, which still is followed by many present-day books on this subject.

5

Gaspard Monge, the director of the Ecole Polytechnique, was the scientific leader of the group of mathematicians who were connected with this institute. He had started his career as instructor at the military academy of Mézières (1768–89), where his lectures on fortification gave him an opportunity to develop descriptive geometry as a special branch of geometry. He published his lectures in the *Géométrie descriptive* (1795–99). At Mézières he also began to apply the calculus to space curves and surfaces, and his papers on this subject were later published in the *Application de l'analyse à la géométrie* (1809), the first book on differential geometry, though not yet in the form that is customary at present. Monge was one of the first modern mathematicians whom we recognize as a specialist: a geometer—even his treatment of partial differential equations has a distinctly geometrical touch.

Through Monge's influence geometry began to flourish at the Ecole Polytechnique. In Monge's descriptive geometry lay the nucleus of projective geometry, and his mastery of algebraic and analytical methods in their application to curves and surfaces contributed greatly to analytical and differential geometry. Jean Hachette and Jean-Baptiste Biot developed the analytical geometry of conics and quadrics; in Biot's *Essai de géométrie analytique* (1802) we begin at last to recognize our present textbooks of analytic geometry. Monge's pupil Charles Dupin, as a young naval engineer in Napoleonic days, applied his teacher's methods to the theory of surfaces, where he found the asymptotic and conjugate lines. Dupin became professor of geometry in Paris, and during his long life gained prominence as a politician and industrial promoter as well. The "indicatrix of Dupin" and the "cyclides of Dupin" remind us of the early interests of this man whose *Développements de géométrie* (1813) and *Applications de géométrie* (1825) contain a great number of interesting results.

The most original pupil of Monge's was Victor Poncelet. He had an opportunity to reflect on his teacher's methods during 1813 when he lived an isolated existence as a prisoner of war in Russia after the defeat of Napoleon's Grande Armée. Poncelet was attracted by the purely synthetic side of Monge's geometry and thus was led to a mode of thinking already suggested two centuries before by Desargues. Poncelet became the founder of projective geometry.

Poncelet's *Traité des propriétés projectives des figures* appeared in 1822. This

GASPARD MONGE (1746–1818)

EVARISTE GALOIS (1811–1832)
(From a rare sketch showing him as he appeared shortly
before his fatal duel at the age of twenty-one.)

heavy volume contains all the essential concepts underlying the new form of geometry, such as cross ratio, perspectivity, projectivity, involution, and even the circular points at infinity. Poncelet knew that the foci of a conic can be considered as the intersections of the tangents at the conic through these circular points. The *Traité* also contains the theory of the polygons inscribed in one conic and circumscribed to another one (the so-called "closure problem" of Poncelet). Although this book was the first full treatise on projective geometry, during the next decades this geometry reached that degree of perfection which made it a classical example of a well-integrated mathematical structure.

6

Monge, although a man of strict democratic principles, stayed loyal to Napoleon, in whom he saw the executor of the ideals of the Revolution. In 1815, when the Bourbons returned, Monge lost his position and soon afterwards died. The Ecole Polytechnique, however, continued to flourish in Monge's spirit. The very nature of the instruction made it difficult to separate pure and applied mathematics. Mechanics received full attention, and mathematical physics began at last to be emancipated from the "catoptrics" and the "dioptrics" of the ancients. Etienne Malus discovered the polarization of light (1810), and Augustin Fresnel re-established Huygens' undulatory theory of light (1821). André-Marie Ampère, who had done distinguished work on partial differential equations, became after 1820 the great pioneer in electromagnetism. These investigators brought many direct and indirect benefits to mathematics: one example is Dupin's improvement of Malus' geometry of light rays, which helped to modernize geometrical optics and also contributed to the geometry of line congruences.

Lagrange's *Mécanique analytique* was faithfully studied and its methods tested and applied. Statics appealed to Monge and his pupils because of its geometrical possibilities, and several textbooks on statics appeared in the course of the years, including one by Monge himself (1788, many editions). The geometrical element in statics was brought out in full by Louis Poinsot, for many years a member of the French superior board of public instruction. His *Eléments de statique* (1804) and *Théorie nouvelle de la rotation des corps* (1834) added to the conception of force that of *torque* ("couple"), represented Euler's theory of moments of inertia by means of the ellipsoid of inertia, and analyzed the motion of this ellipsoid when the rigid body moves in space or turns about a fixed point. Poncelet and Coriolis gave a geometrical touch to Lagrange's analytical mechanics; both men, as well as Poinsot, stressed the application of mechanics to the theory of simple machines. The "acceleration of Coriolis," which appears when a body moves in an accelerated system, is an example of such a geometrical interpretation of Lagrange's results (1835).

The most outstanding mathematicians connected with the early years of the Ecole Polytechnique were—apart from Lagrange and Monge—Siméon Poisson, Joseph Fourier, and Augustin Cauchy. All three were deeply interested in the application of mathematics to mechanics and physics, and all three were led by this interest to discoveries in "pure" mathematics. Poisson's productivity is indicated by the frequency with which his name occurs in our textbooks: "Poisson's brackets" in differential equations, "Poisson's constant" in elasticity, "Poisson's integral," and "Poisson's equation" in potential theory. This "Poisson equation," $\Delta V = 4\,\pi\rho$, was the result of Poisson's discovery (1812) that Laplace's equation, $\Delta V = 0$, only holds outside of the masses; its exact proof for masses of variable density was not given until Gauss gave it in his *Allgemeine Lehrsätze* (1839–40). Poisson's *Traité de mécanique* (1811) was written in the spirit of Lagrange and Laplace but contained many innovations, such as the explicit use of impulse coordinates, $p_i = \partial T/\partial \dot{q}_i$, which later inspired the work of Hamilton and Jacobi. His book of 1837 contains "Poisson's law" in the theory of probabilities, originally derived as an approximation to the binomial law of Bernoulli for the case of small probabilities, but now recognized as a fundamental law in problems of radiation, traffic, and distribution in general (see Chap. VII, Sec. 6, above).

Fourier is primarily remembered as the author of the *Théorie analytique de la chaleur* (1822).[5] This is the mathematical theory of heat conduction and, therefore, is essentially the study of the equation $\Delta U = k\partial U/\partial t$. By virtue of the generality of its method this book became the source of all modern methods in mathematical physics involving the integration of partial differential equations under given boundary conditions. This method is the use of trigonometric series, which had been the cause of discussion between Euler, d'Alembert, and Daniel Bernoulli. Fourier made the situation perfectly clear. He established the fact that an "arbitrary" function (a function capable of being represented by an arc of a continuous curve or by a succession of such arcs) could be represented by a trigonometric series of the form $\sum_{n=0}^{\infty}(A_n \cos nax + B_n \sin nax)$. Despite Euler's and Bernoulli's observations, the idea was so new and startling at the time of Fourier's investigations that it is said that, when he stated his ideas in 1807 for the first time, he met with the vigorous opposition of none other than Lagrange himself.

The "Fourier series" now became a well-established instrument of operation in the theory of partial differential equations with given boundary conditions. It also received attention on its own merits. Its manipulation by Fourier fully opened the question of what to understand by a "function." This was one of the reasons why nineteenth-century mathematicians found it necessary to look more closely into questions concerning rigor of mathematical proof and the foundation of mathematical conceptions in general.[6] This task, in the specific case of Fourier series, was undertaken by Dirichlet and Riemann.

7

Cauchy's many contributions to the theory of light and to mechanics have been obscured by the success of his work in analysis, but we must not forget that with Navier he belongs to the founders of the mathematical theory of elasticity. His

---

[5]Translated as *The Analytic Theory of Heat* by Alexander Freeman.

[6]P. E. B. Jourdain, "Note on Fourier's Influence on the Conceptions of Mathematics," *Proc. Intern. Congress of Math.*, Vol. II (Cambridge, 1912), pp. 526–27.

main glory is the theory of functions of a complex variable and his insistence on rigor in analysis. Functions of a complex variable had already been constructed before, notably by d'Alembert, who in a paper on the resistance of fluids of 1752 had even obtained what we now call the Cauchy-Riemann equations. In Cauchy's hands complex function theory was emancipated from a tool useful in hydrodynamics and aerodynamics into a new and independent field of mathematical research. Cauchy's investigations on this subject appeared in constant succession after 1814. One of the most important of his papers is the *Mémoire sur les intégrales définies, prises entre des limites imaginaires* (1825). In this paper appeared Cauchy's integral theorem with residues. The theorem that every regular function $f(z)$ can be expanded around each point $z = z_0$ in a series convergent in a circle passing through the singular point nearest to $z = z_0$ was published in 1831, the same year in which Gauss published his arithmetical theory of complex numbers. Laurent's extension of Cauchy's theorem on series was published in 1843— when it was also in the possession of Weierstrass. These facts show that Cauchy's theory did not have to cope with professional resistance; the theory of complex functions was fully accepted from its very beginning.

Cauchy, along with his contemporaries—Gauss, Abel, and Bolzano—belongs to the pioneers of the new insistence on rigor in mathematics. The eighteenth century had been essentially a period of experimentation, in which results came pouring in with luxurious abundance. The mathematicians of this age had not paid too much attention to the foundation of their work—*"allez en avant, et la foi vous viendra,"*[7] d'Alembert is supposed to have said. When they worried about rigor, as Euler and Lagrange occasionally did, their arguments were not always convincing. The time had now arrived for a close concentration on the meaning of the results. What was a "function" of a real variable, which showed such different behavior in the case of a Fourier series and in the case of a power series? What was its relation to the entirely different "function" of a complex variable? These questions brought all the unsolved problems about the foundation of the calculus and the existence of the potentially and the actually infinite back into the foreground of mathematical thinking.[8] What Eudoxus had done in the period after the fall of Athenian democracy, Cauchy and his meticulous contemporaries began to accomplish in the period of expanding industrialism. This difference of social setting produced different results, and where Eudoxus' success had the tendency to stifle productivity, the success of the modern reformers stimulated mathematical productivity to a high degree. Cauchy and Gauss were followed by Weierstrass and Cantor.

Cauchy gave the foundation of the calculus as we now generally accept it in our textbooks. It can be found in his *Cours d'analyse* (1821), and *Résumé des leçons données à l'Ecole Royale Polytechnique* (Vol. I, 1823). Cauchy used d'Alembert's limit concept to define the derivative of a function, and establish it on a firmer foundation than his predecessors had been able to do.

---

[7]"Go forward and faith will come to you."

[8]P. E. B. Jourdain, "The Origin of Cauchy's Conception of a Definite Integral and of the Continuity of a Function," *Isis*, Vol. 1 (1913), pp. 661–703 (see also *Bibl. Math.*, Vol. 6 [1905], pp. 190–207).

Starting with the definition of a "limit," Cauchy gave examples such as the limit of sin α/α for $\alpha = 0$. Then he defined a *variable infiniment petite* as a variable number that has zero for its limit; and then postulated that Δy and Δx *seront des quantités infiniment petites*. He then wrote

$$\frac{\Delta y}{\Delta x} = \frac{f(x + i) - f(x)}{i}$$

and called the limit for $i \to 0$ the *fonction dérivée*, y' ou $f'(x)$. He placed $i = \alpha h$, α an *infiniment petite*, and h a *quantité finie*:

$$\frac{f(x + \alpha h) - f(x)}{\alpha} = \frac{f(x + i) - f(x)}{i} h$$

h was called the *différentielle de la fonction y = f(x)*. Furthermore, $dy = df(x) = hf'(x)$; $dx = h$.[9]

Cauchy used both Lagrange's notation and many of his contributions to real function theory without making any concession to Lagrange's "algebraic" foundation. The mean-value theorem and the remainder of the Taylor series were accepted as Lagrange had derived them, but series were now discussed with due attention to their convergence. Several convergence proofs in the theory of infinite series are named after Cauchy. We find in his works our modern concept of continuity. There are definite steps in his books toward that "arithmetization" of analysis which later became the core of Weierstrass' investigations. Cauchy also gave the first existence proof for the solution of a differential equation and of a system of such equations (1836). In this way Cauchy offered at last a beginning of an answer to that series of problems and paradoxes which had haunted mathematics since the days of Zeno, and he did this not by denying or ignoring them, but by creating a mathematical technique with which it was possible to do them justice.[10]

Cauchy, like his contemporary Balzac (with whom he shared a capacity for an infinite amount of work), was a legitimist and a royalist. Both men had such a deep understanding of values that despite their reactionary ideals much of their work has retained its fundamental place. Cauchy abandoned his chair at the École Polytechnique after the Revolution of 1830 and spent some years at Turin and Prague; he returned to Paris in 1838. After 1848 he was allowed to stay and teach without having to take the oath of allegiance to the new government. His productivity was so enormous that the French Academy had to restrict the size of all papers sent to the *Comptes-Rendus* in order to cope with Cauchy's output. It is told that he disturbed Laplace so greatly when he read to the French Academy his first paper on the convergence of series that the great man went back home to test the

---

[9]*Résumé*, Vol. I (1823), Chapter entitled "Calcul différentiel," pp. 13–27. A strict analysis of this procedure is given in M. Pasch, *Mathematik am Ursprung* (Leipzig, 1927), pp. 47–73.

[10]See, e.g., Chap. 40, "The Installation of Rigor in Analysis," in M. Kline, *Mathematical Thought from Ancient to Modern Times* (New York, 1972); also H. Freudenthal, *DSB*, Vol. 3 (1971), pp. 131–48.

series in his *Mécanique céleste*. He found, it seems, that no great errors had been committed.

8

This Paris milieu with its intense mathematical activity produced, around 1830, a genius of the first order, who, comet-like, disappeared as suddenly as he had appeared. Evariste Galois, the son of a small-town mayor near Paris, was refused admission to the Ecole Polytechnique but succeeded at last in entering the Ecole Normale only to be dismissed. He felt oppressed by the representatives of official science; one of the papers he sent for publication to the Academy was returned to him by Cauchy, and his new text, written in the hope of getting the *grand prix*, was lost. His ardent spirit tried to maintain an uneasy balance between his aversion to the "establishment" and his passion for science and democracy. Galois participated as a republican in the Revolution of 1830, spent some months in prison, and was soon afterwards killed in a duel at the age of twenty-one. Some of his papers were only published long after his death. On the eve of the duel, he wrote to a friend a summary of his discoveries in the theory of equations. This pathetic document, in which he asked his friend to submit his discoveries to the leading mathematicians, ended with the words:

> You will publicly ask Jacobi or Gauss to give their opinion not on the truth, but on the importance of the theorems. After this there will be, I hope, some people who will find it to their advantage to decipher all this mess.

This "mess" (*gâchis*) contained no less than the theory of groups, the key to modern algebra and to modern geometry. The ideas had already been anticipated to a certain extent by Lagrange and the Italian Ruffini, but Galois had the conception of a complete theory of groups. He expressed the fundamental properties of the transformation group belonging to the roots of an algebraic equation and showed that the field of rationality of these roots was determined by the group. Galois pointed out the central position taken by invariant subgroups. Ancient problems such as the trisection of the angle, the duplication of the cube, and the solution of the cubic and biquadratic equation, as well as the solution of an algebraic equation of any degree, found their natural place in the theory of Galois. Galois's letter, as far as we know, was never submitted to Gauss or Jacobi. It never reached a mathematical public until Liouville published most of Galois's papers in his *Journal de mathématiques* of 1846, at which period Cauchy had already begun to publish on group theory (1844–46). It was only then that some mathematicians began to be interested in Galois's theories. Full understanding of Galois's importance came only through Camille Jordan's *Traité des substitutions* (1870) and the subsequent publications by Klein and Lie. Now Galois's unifying principle has been recognized as one of the outstanding achievements of nineteenth-century mathematics.[11]

---

[11]See G. A. Miller, *History of the Theory of Groups to 1900. Collected Works*, Vol. I (1935), pp. 427–67; H. Wussing, *Die Genesis des abstrakten Gruppenbegriffes* (Berlin, 1969); on Jordan, see H. Lebesgue, *Notices d'histoire des mathématiques* (Geneva, 1958), pp. 40–65.

Galois also had ideas on the integrals of algebraic functions of one variable, on what we now call Abelian integrals. This brings his manner of thinking closer to that of Riemann. We may speculate on the possibility that if Galois had lived, modern mathematics might have received its deepest inspiration from Paris and the school of Lagrange rather than from Göttingen and the school of Gauss.

9

Another young genius appeared in the twenties, Niels Henrik Abel, the son of a Norwegian country minister. Abel's short life was almost as tragic as that of Galois. As a student in Christiania he thought for a while that he had solved the equation of degree five, but he corrected himself in a pamphlet published in 1824. This was the famous paper in which Abel proved the impossibility of solving the general quintic equation by means of radicals—a problem which had puzzled mathematicians from the time of Bombelli and Viète (a proof of 1799 by the Italian Paolo Ruffini was considered by Poisson and other mathematicians as too vague). Abel now obtained a stipend which enabled him to travel to Berlin, Italy, and France. But, tortured by poverty most of his life and unable to get a position worthy of his talents, Abel established few personal mathematical contacts and died (1829) soon after his return to his native land, at a moment when his greatness began to be recognized by Legendre, Jacobi, and Gauss.

In this period of travel Abel wrote several papers which contain his work on the convergence of series, on "Abelian integrals," and on elliptic functions. Abel's theorems in the theory of infinite series show that he was able to establish this theory on a reliable foundation. "Can you imagine anything more horrible than to claim that $0 = 1^n - 2^n + 3^n - 4^n +$ etc., n being a positive integer?" he wrote to a friend, and he continued:

> There is in mathematics hardly a single infinite series of which the sum is determined in a rigorous way (Letter to Holmboe, 1826).

Abel's investigations on elliptic functions were conducted in a short but exciting competition with Jacobi. Gauss in his private notes had already found that the inversion of elliptic integrals leads to single-valued, doubly periodic functions, but he never published his ideas. Legendre, who had spent so much effort on elliptic integrals, had missed this point entirely and was deeply impressed when, as an old man, he read Abel's discoveries. Abel had the good luck to find a new periodical eager to print his papers; the first volume of the *Journal für die reine und angewandte Mathematik*, edited by Crelle, contained no less than five of Abel's papers. In the second volume (1827) appeared the first part of his *Recherches sur les fonctions elliptiques*, with which the theory of doubly periodic functions begins.

August Leopold Crelle, architect and planner of the first railroad in Prussia, is still known through his *Rechentafeln* (1820) and as the founder of the still existing *Journal für die reine und angewandte Mathematik* (1826), one of the first purely mathematical journals, where Crelle encouraged young mathematicians such as

Abel and Steiner. It took its name from Gergonne's *Annales de mathématiques pures et appliquées* (after 1836, *Journal* . . .; see Sec. 16, below).

We speak of Abel's integral equation and of Abel's theorem on the sum of integrals of algebraic functions which leads to Abelian functions. Commutative groups are called Abelian groups, which indicates how closely Galois's ideas were related to those of Abel.

10

In 1829, the year in which Abel died, Carl Gustav Jacob Jacobi published his *Fundamenta nova theoriae functionum ellipticarum*. The author was a young professor at the University of Königsberg. He was the son of a Berlin banker and a member of a distinguished family; his brother Moritz in St. Petersburg was one of the earliest Russian scientists who experimented in electricity. After studying in Berlin, Jacobi taught at Königsberg from 1826 to 1843. He then spent some time in Italy trying to regain his health and ended his career as a professor at the University of Berlin, dying in 1851 at the age of forty-six. He was a witty and liberal thinker, an inspiring teacher, and a scientist whose enormous energy and clarity of thought left few branches of mathematics untouched.

Jacobi based his theory of elliptic functions on four functions defined by infinite series and called "Theta functions." The doubly periodic functions $sn\ u$, $cn\ u$, and $dn\ u$ are quotients of Theta functions; they satisfy certain identities and addition theorems very much like the sine and cosine functions of ordinary trigonometry. The addition theorems of elliptic functions can also be considered as special applications of Abel's theorem on the sum of integrals of algebraic functions. The question now arose whether hyperelliptic integrals could be inverted in the way elliptic integrals had been inverted to yield elliptic functions. The solution was found by Jacobi in 1832 when he published his result that the inversion could be performed with functions of more than one variable. Thus was born the theory of Abelian functions of p variables, which became an important branch of nineteenth-century mathematics.

Sylvester has given the name "Jacobian" to the functional determinant in order to pay respect to Jacobi's work on algebra and elimination theory. The best-known of Jacobi's papers on this subject is his *De formatione et proprietatibus determinantium* (1841), which made the theory of determinants the common property of the mathematicians. The idea of the determinant is much older—it dates back essentially to Leibniz (1693), the Swiss mathematician Gabriel Cramer (1750), and Lagrange (1773); the name is due to Cauchy (1812). Y. Mikami has pointed out that the Japanese mathematician Seki Kōwa had the idea of a determinant sometime before 1683.[12] We are here reminded of the "matrix" methods developed by the Chinese mathematicians of the Song period, whose works Seki had studied.

---

[12]Y. Mikami, "On the Japanese Theory of Determinants," *Isis*, Vol. 2 (1914), pp. 9–36. According to J. Needham (*Science and Civilization in China*, Vol. III [Cambridge, 1959], p. 117), a better reading of the name is Seki Takakusu. He wrote on the numerical solution of algebraic equations of degree n and the rectification of the circular arc. The *DSB* article spells the name Seki Takakazu. The influence of the Chinese tradition in his mathematics is clear.

The best approach to Jacobi is perhaps through his beautiful lectures on dynamics (*Vorlesungen über Dynamik*), published in 1866 after lecture notes of 1842–43. They are written in the tradition of the French school of Lagrange and Poisson but have a wealth of new ideas. Here we find Jacobi's investigations on partial differential equations of the first order and their application to the differential equations of dynamics. An interesting chapter of the *Vorlesungen über Dynamik* is the determination of the geodesics on an ellipsoid; the problem leads to a relation between two Abelian integrals.

11

Jacobi's lectures on dynamics lead us to another mathematician whose name is often linked with that of Jacobi, William Rowan Hamilton (not to be confused with his contemporary, William Hamilton, the Edinburgh philosopher). He lived his whole life in Dublin, where he was born of Irish parents. He entered Trinity College, became Royal Astronomer of Ireland in 1827 at the age of twenty-one, and held this position until his death in 1865. As a boy he learned continental mathematics—still a novelty in the United Kingdom—by studying Clairaut and Laplace, and showed his mastery of the novel methods in his extremely original work on optics and dynamics. His theory of optical rays (1824) was more than merely a differential geometry of line congruences; it was also a theory of optical instruments and allowed Hamilton to predict the conical refractions in biaxial crystals. In this paper appeared his "characteristic function," which became the guiding idea of the *General Method in Dynamics*, published in 1834–35. Hamilton's idea was to derive both optics and dynamics from one general principle. Euler, in his defense of Maupertuis, had already shown how the stationary value of the action integral could serve this purpose. Hamilton, in accordance with this suggestion, made optics and dynamics two aspects of the calculus of variations. He asked for the stationary value of a certain integral and considered it as a function of its limits. This was the "characteristic" or "principal" function, which satisfied two partial differential equations. One of these partial differential equations, which is usually written

$$\frac{\partial S}{\partial t} + H\left(\frac{\partial S}{\partial q}, q\right) = 0,$$

was specially selected by Jacobi for his lectures on dynamics and is now known as the Hamilton-Jacobi equation. This has obscured the importance of Hamilton's characteristic function, which had the central place in his theory as a means of unifying mechanics and mathematical physics. It was rediscovered by the Leipzig astronomer Heinrich Bruns in 1895 in the case of geometrical optics and as *eikonal* has shown its use in the theory of optical instruments.

The part of Hamilton's work on dynamics which has passed into the general body of mathematics is, in the first place, the "canonical" form, $\dot{q} = \partial H/\partial p$, $\dot{p} = -\partial H/\partial q$, in which he wrote the equations of dynamics. Canonical form and the Hamilton-Jacobi differential equation enabled Lie to establish the relation between dynamics and contact transformations. Another idea of Hamilton, since equally accepted, was the derivation of the laws of physics and mechanics from the variation of an integral. Modern relativity, as well as quantum mechanics, has based itself on "Hamiltonian functions" as its underlying principle.

The year 1843 was a turning point in Hamilton's life. In that year he found the quaternions, to the study of which he devoted the latter part of his life. We shall discuss this discovery later.

12

Peter Lejeune Dirichlet was closely associated with Gauss and Jacobi, as well as with the French mathematicians. He lived from 1822 to 1827 as a private tutor and met Fourier, whose book he studied; he also became familiar with Gauss's *Disquisitiones arithmeticae*. He later taught at the University of Breslau and in 1855 succeeded Gauss at Göttingen. His personal acquaintance with French as well as German mathematics and mathematicians made him the appropriate man to serve as the interpreter of Gauss and to subject Fourier series to a penetrating analysis. Dirichlet's beautiful *Vorlesungen über Zahlentheorie* (1863) still forms one of the best introductions to Gauss's investigations in number theory. It also contains many new results. In a paper of 1840 Dirichlet showed how to apply the full power of the theory of analytical functions to problems in number theory; it was in these investigations that he introduced the "Dirichlet series." He also extended the notion of quadratic irrationalities to that of general algebraic domains of rationality.

Dirichlet first gave a rigorous convergence proof of Fourier series, and in this way contributed to a correct understanding of the nature of a function.[13] He also introduced into the calculus of variations the so-called Dirichlet principle, which postulated the existence of a function v which minimizes the integral $\int [v_x^2 + v_y^2 + v_z^2] \, d\tau$ under given boundary conditions. It was a modification of a principle which Gauss had introduced in his potential theory of 1839–40, and it later served Riemann as a powerful tool in solving problems in potential theory. We have already mentioned that eventually Hilbert rigorously established the validity of this principle (Sec. 2, above).[14]

13

With Bernhard Riemann, Dirichlet's successor at Göttingen, we reach the man who more than any other influenced the course of modern mathematics. Riemann was the son of a country minister and studied at the University of Göttingen, where in 1851 he obtained the doctor's degree. In 1854 he became Privatdozent, in 1859 a professor at the same university. He was sickly, like Abel, and spent his last days in Italy, where he died in 1866 at forty years of age. In his short life he

---

[13]See A. F. Monna, "The Concept of Function in the 19th and 20th Centuries, in Particular with Regard to the Discussions Between Baire, Borel and Lebesgue," *AHES*, Vol. 9 (1972), pp. 51–84.

[14]A. F. Monna, *Dirichlet's Principle* (Utrecht, 1975).

published only a relatively small number of papers but each of them was—and is—important, and several have opened entirely new and productive fields.

In 1851 appeared Riemann's doctoral thesis on the theory of complex functions $u + iv = f(x + iy)$. Like d'Alembert and Cauchy, Riemann was influenced by hydrodynamical considerations. He mapped the (xy)-plane conformally on the (uv)-plane and established the existence of a function capable of transforming any simply connected region in one plane into any simply connected region in the other plane. This led to the conception of the Riemann surface, which introduced topological considerations into analysis. At that time topology was still an almost untouched subject, on which J. B. Listing had published a paper in the *Göttinger Studien* of 1847. Riemann showed its central importance for the theory of complex functions. This thesis also clarified Riemann's definition of a complex function: its real and imaginary parts have to satisfy the "Cauchy-Riemann equations," $u_x = v_y$, $u_y = -v_x$, in a given region, and furthermore have to satisfy certain conditions as to boundary and singularities.

Riemann applied his ideas to hypergeometric and Abelian functions (1857), freely using Dirichlet's principle (as he called it). Among his results was the discovery of the genus of a Riemann surface as a topological invariant and as a means of classifying Abelian functions. A posthumously published paper applies his ideas to minimal surfaces (1867). To this branch of Riemann's activity also belong his investigations on elliptical modular-functions and θ-series in p independent variables, as well as those on linear differential equations with algebraic coefficients.

Riemann became a Privatdozent in 1850 by submitting no less than two fundamental papers, one on trigonometric series and the foundations of analysis, the other on the foundations of geometry. The first of these papers analyzed Dirichlet's conditions for the expansion of a function in a Fourier series. One of these conditions was that the function be "integrable." But what does this mean? Cauchy and Dirichlet had already given certain answers; Riemann replaced them by his own, more comprehensive one. He gave that definition which we now know as the "Riemann integral," and which was replaced only in the twentieth century by the Lebesgue integral. Riemann showed how functions, defined by Fourier series, may exhibit such properties as the possession of an infinite number of maxima or minima, which older mathematicians would not have accepted in their definition of a function. The concept of a function began seriously to be emancipated from the *curva quaecunque libero manus ductu descripta*[15] of Euler. In his lectures Riemann gave an example of a continuous function without derivatives; an example of such a function which Weierstrass had given was published in 1875. Mathematicians refused to take such functions very seriously and called them "pathological" functions; modern analysis has shown how natural such functions are and how Riemann here again had penetrated into a fundamental field of mathematics.

The other paper of 1854 deals with the hypotheses on which geometry is based. Space was introduced as a topological manifold of an arbitrary number of

---

[15]"Some curve described by freely leading the hand" (*Inst. Calc. integr.*, Vol. III, §301). See footnote 15, Chap. VII, above.

GEORG FRIEDRICH BERNHARD RIEMANN
(1826–1866)

KARL WEIERSTRASS (1815–1897)

dimensions; a metric was defined in such a manifold by means of a quadratic differential form. Where Riemann, in his analysis, had defined a complex function by its local behavior, in this paper he defined the character of space in the same way. Riemann's unifying principle not only enabled him to classify all existing forms of geometry, including the still very obscure non-Euclidean geometry, but also allowed the creation of any number of new types of space, many of which have since found a useful place in geometry and mathematical physics. Riemann published this paper without any analytical technique, which made his ideas difficult to follow. Later some of the formulas appeared in a prize essay on the distribution of heat in a solid, which Riemann submitted to the French Academy (1861). Here we find a sketch of the transformation theory of quadratic forms.

The last paper of Riemann which must be mentioned is his discussion of the number $F(x)$ of primes less than a given number x (1859). It was an application of complex number theory to the distribution of primes, and it analyzed Gauss's suggestion that $F(x)$ approximates the logarithmic integral $\int_2^x (\log t)^{-1} dt$. This paper is celebrated because it contains the so-called Riemann hypothesis that Euler's Zeta function $\zeta(s)$—the notation is Riemann's—if considered for complex $s = x + iy$, has all nonreal zeros on the line $x = \frac{1}{2}$. This hypothesis has never been proved, nor has it been disproved.[16]

14

Riemann's concept of the function of a complex variable has often been compared to that of Weierstrass. Karl Weierstrass was for many years a teacher at Prussian gymnasia (Latin high schools), and in 1856 became professor of mathematics at the University of Berlin, where he taught for thirty years. His lectures, always meticulously prepared, enjoyed increasing fame; it is mainly through these lectures that Weierstrass' ideas have become the common property of mathematicians.

In his gymnasial period Weierstrass wrote several papers on hyperelliptic integrals, Abelian functions, and algebraic differential equations. His best-known contribution is his foundation of the theory of complex functions on the power series. This was in a certain sense a return to Lagrange, with the difference that Weierstrass worked in the complex plane and with perfect rigor. The values of the power series inside its circle of convergence represent the "function element," which is then extended, if possible, by so-called analytic continuation. Weierstrass especially studied entire functions defined by infinite products. His elliptic function $\wp(u)$ has become as established as the older $sn\ u$, $cn\ u$, $dn\ u$ of Jacobi.

Weierstrass' fame has been based on his extremely careful reasoning, on "Weierstrassian rigor," which is not only apparent in his function theory but also in his calculus of variations. He clarified the notions of minimum, of function, and of derivative, and with this eliminated the remaining vagueness of expression in

---

[16]R. Courant, "Bernhard Riemann und die Mathematik der letzten hundert Jahre," *Naturwissenschaften*, Vol. 14 (1926), pp. 813–18; E. C. Titchmarsh, *The Theory of the Riemann Zeta-Function* (Oxford, 1951); H. H. Edwards, *Riemann's Zeta Function* (New York, 1974).

the fundamental concepts of the calculus. He was mathematical conscientiousness par excellence, methodological and logical. Another example of his meticulous reasoning is his discovery of uniform convergence. With Weierstrass began that reduction of the principles of analysis to the simplest arithmetical concepts which we call the arithmetization of mathematics.

It is essentially a merit of the scientific activity of Weierstrass that there exists at present in analysis full agreement and certainty concerning the course of such types of reasoning which are based on the concept of irrational number and of limit in general. We owe it to him that there is unanimity on all results in the most complicated questions concerning the theory of differential and integral equations, despite the most daring and diversified combinations with application of super-, juxta-, and transposition of limits.[17]

15

This arithmetization was typical of the so-called School of Berlin and especially of Leopold Kronecker. To this school belonged eminent mathematicians, proficient in algebra and the theory of algebraic numbers, such as Kronecker, Kummer, and Frobenius. With these men we may associate Dedekind and Cantor. Ernst Kummer was called to Berlin in 1855 as successor to Dirichlet; he taught there until 1883, when he voluntarily stopped doing mathematical work because he felt the onset of a decline in productivity. Kummer further developed the differential geometry of congruences which Hamilton had outlined and in the course of his study discovered the quartic surface with sixteen nodal points that is named after him. His reputation is primarily based on his introduction of the "ideal" numbers in the theory of algebraic domains of rationality (1846). This theory was inspired partly by Kummer's attempts to prove Fermat's great theorem and partly by Gauss's theory of biquadratic residues in which the conception of prime factors had been introduced within the domain of complex numbers. Kummer's "ideal" factors allowed a unique decomposition of numbers into prime factors in a general domain of rationality. This discovery made possible great advances in the arithmetic of algebraic numbers, which were later masterfully summarized in the report of David Hilbert written for the German Mathematical Society in 1894.[18] The theory of Dedekind and Weber, which established a relation between the theory of algebraic functions and the theory of algebraic numbers in a certain domain of rationality (1882), was an example of the influence of Kummer's theory on the process of arithmetization of mathematics.

Leopold Kronecker, a man of private means, settled in Berlin in 1855, where he taught for many years at the university without a formal professional chair, only accepting one after Kummer's retirement in 1883. Kronecker's main contributions were in elliptic functions, ideal theory, and in the arithmetic of quadratic forms; his published lectures on the theory of numbers are careful expositions of his own and of previous discoveries, and also show clearly his belief in the necessity of the

[17]D. Hilbert, "Über das Unendliche," *Mathematische Annalen*, Vol. 95 (1926), pp. 161–90; French translation, *Acta Mathematica*, Vol. 48 (1926), pp. 91–122.
[18]D. Hilbert, "Die Theorie der algebraischen Zahlkörper," in *Jahresber. Deut. Math. Verein.*, 4 (1894–95), pp. 175–546.

arithmetization of mathematics. This belief was based on his search for rigor; mathematics, he thought, should be based on number, and all number on the natural number. The number π, for instance, rather than be derived in the usual geometrical way, should be based on the series $1 - \frac{1}{3} + \frac{1}{5} - \frac{1}{7} + \cdots$ and thus on a combination of integer numbers; certain continued fractions for π might also serve the same purpose. Kronecker's endeavor to force everything mathematical into the pattern of number theory is illustrated by his well-known statement at a meeting in Berlin in 1886: *"Die ganzen Zahlen hat der liebe Gott gemacht, alles andere ist Menschenwerk."*[19] He accepted a definition of a mathematical entity only when it could be verified in a finite number of steps. Thus he coped with the difficulty of the actually infinite by refusing to accept it. Plato's slogan that God always "geometrizes" was replaced, in Kronecker's school, by the slogan that God always "arithmetizes."

Kronecker's teaching on the actually infinite was in flagrant contrast to the theories of Dedekind and especially of Cantor. Richard Dedekind, for thirty-one years professor at the Technische Hochschule in Brunswick, constructed a rigorous theory of the irrational. In two small books, *Stetigkeit und irrationale Zahlen* (1872) and *Was sind und was sollen die Zahlen?* (1888),[20] he accomplished for modern mathematics what Eudoxus had done for Greek mathematics. There is a great similarity between the "Dedekind cut" with which modern mathematics (except the Kronecker school) defines irrational numbers and the ancient Eudoxus theory as presented in the fifth book of Euclid's *Elements*. Cantor and Weierstrass gave arithmetical definitions of the irrational numbers differing somewhat from the Dedekind theory but based on similar considerations.

The greatest heretic in Kronecker's eyes, however, was Georg Cantor. Cantor, who taught at Halle from 1869 until 1905, is known not only because of his theory of the irrational number, but also because of his theory of aggregates (*Mengenlehre*). With this theory Cantor created an entirely new field of mathematical research, which was able to satisfy the most subtle demands of rigor once its premises were accepted. Cantor's publications began in 1870 and continued for many years; in 1883 he published his *Grundlagen einer allgemeinen Mannigfaltigkeitslehre*. In these papers Cantor developed a theory of transfinite cardinal numbers based on a systematical mathematical treatment of the actually infinite. He assigned the lowest transfinite cardinal number $\aleph$ to a denumerable set, giving the continuum a higher transfinite number, and it thus became possible to create an arithmetic of transfinite numbers generalizing ordinary arithmetic. Cantor also defined transfinite ordinal numbers, expressing the way in which infinite sets are ordered.

These discoveries of Cantor were a continuation of the ancient scholastic speculations on the nature of the infinite, and Cantor was well aware of it. He defended St. Augustine's full acceptance of the actually infinite, but had to defend

---

[19]"The integer numbers were made by God, everything else is the work of man."

[20]Translated as "Continuity and Irrational Numbers" and "The Nature and Meaning of Numbers" by W. W. Beman (Chicago, 1901); reprinted together as *Essays on the Theory of Numbers*, Dover Publications, Inc., 1963.

GEORG CANTOR (1845–1918)

himself against the opposition of many mathematicians who refused to accept the infinite except as a process expressed by ∞. Cantor's leading opponent was Kronecker, who represented a totally opposite tendency in the same process of arithmetization of mathematics. Cantor finally won broad acceptance when the enormous importance of his theory for the foundation of real function theory and of topology became more and more obvious—this especially after Lebesgue in 1901 had enriched the theory of aggregates with his theory of measure. There remained logical difficulties in the theory of transfinite numbers and paradoxes appeared, such as those of Burali Forti and Russell. This again led to different schools of thought on the foundation of mathematics. The twentieth-century controversy between the formalists and the intuitionists was a continuation on a novel level of the controversy between Cantor and Kronecker.

16

Contemporaneous with this remarkable development of algebra and analysis was the equally remarkable flowering of geometry. It can be traced back to Monge's instruction, in which we find the roots of both the "synthetic" and the "algebraic" method in geometry. In the work of Monge's pupils both methods became separated: the "synthetic" method developed into projective geometry, and the "algebraic" method into our modern analytical and algebraic geometry. Projective geometry as a separate science began with Poncelet's book of 1822. There were priority difficulties, as so often in cases concerning a fundamental discovery, since Poncelet had to face the rivalry of Joseph Gergonne, professor at Montpellier. Gergonne published several important papers on projective geometry in which he grasped the meaning of duality in geometry simultaneously with Poncelet. These papers appeared in the *Annales de mathématiques pures et appliquées*, the first purely mathematical periodical. Gergonne was its editor; it appeared from 1810 to 1831. (In 1836 it resumed publication under the name *Journal de mathématiques pures et appliquées*.)

Typical of Poncelet's mode of thinking was another principle, that of continuity, which enabled him to derive the properties of one figure from those of another. He expressed the principle as follows:

> Consider an arbitrary figure in a general position and in some respects undetermined among all those positions which it can take without violating the laws, the conditions, the connection existing between the different parts of the system. Suppose that one or more relations or properties—metrical ones, descriptive ones—have been found in accordance with these data. . . . Is it not evident that when, conserving these same data, we subject certain parts of the original figure to some continuous movement, that the properties and relations that were found for the first system will be applicable to the successive stages of this system, if only the particular modifications are taken into account . . . for instance, certain quantities may have vanished or changed their sense or sign. . . .[21]

This was a principle which had to be handled with great care, since the formulation was far from precise. Only modern algebra has been able to define its domain more accurately. In the hands of Poncelet and his school it led to interesting, new, and accurate results, especially when it was applied to changes from the real to the imaginary. It thus enabled Poncelet to state that all circles in the plane had "ideally two imaginary points at infinity in common," which also brought in the so-called "line at infinity" of the plane. G. H. Hardy has remarked that this meant that projective geometry accepted the actually infinite without any scruples.[22] The analysts were to remain divided on this subject.

Poncelet's ideas were further developed by German geometers. In 1826 appeared the first of Steiner's publications, in 1827 Möbius' *Barycentrischer Calcül*, in 1828 the first volume of Plücker's *Analytisch-geometrische Entwickelungen*. In

---

[21]Poncelet, *Traité des propriétés projectives des figures* (1882), p. xiii.
[22]G. H. Hardy, *A Course of Pure Mathematics*, 6th ed. (Cambridge, 1933), Appendix IV.

1831 appeared the second volume, followed in 1832 by Steiner's *Systematische Entwickelung*. The last of the great German pioneer works in this type of geometry appeared in 1847 with the publication of von Staudt's axiomatic *Geometrie der Lage*.

Both the synthetic and the algebraic approaches to geometry were represented among these German geometers. The typical representative of the synthetic (or "pure") school was Jakob Steiner, a self-educated Swiss farmer's son, a *Hirtenknabe*, who became enamored of geometry by making the acquaintance of Pestalozzi's ideas. He decided to study at Heidelberg and later taught at Berlin, where from 1834 until his death in 1863 he held a chair at the university. Steiner was thoroughly a geometer; he hated the use of algebra and analysis to such an extent that he even disliked figures. Geometry in his opinion could best be learned by concentrated thought. Calculating, he said, replaces, while geometry stimulates, thinking. This was certainly true for Steiner himself, whose methods have enriched geometry with a large number of beautiful and often intricate theorems. We owe to him the discovery of the Steiner surface with a double infinity of conics on it (also called the Roman surface). He often omitted the proof of his theorems, which has made Steiner's collected works a treasure trove for geometers in search of problems to solve.

Steiner constructed his projective geometry in a strictly systematic way, passing from perspectivity to projectivity and from there to the conic sections. He also solved a number of isoperimetrical problems in his own typical geometrical way. His proof (1836) that the circle is the figure of largest area for all closed curves of given perimeter made use of a procedure by which every figure of given perimeter which is not a circle could be changed into another one of the same perimeter and of larger area. Steiner's conclusion that the circle therefore represented the maximum suffered from one omission: he did not prove the actual existence of a maximum. Dirichlet tried to point it out to Steiner; a rigorous proof was later given by Weierstrass.[23]

Steiner still needed a metric to define the cross ratio of four points or lines. This defect in the theory was removed by Christian von Staudt, for many years a professor at the University of Erlangen. Von Staudt, in his *Geometrie der Lage*, defined the *Wurf* of four points on a straight line in a purely projective way, and then showed its identity with the cross ratio. He used for this purpose the so-called Möbius net construction, which leads to axiomatic considerations closely related to Dedekind's work when irrational values of projective coordinates are introduced. In 1857 von Staudt showed how imaginary elements can be rigorously introduced into geometry as double elements of elliptic involutions.

During the next decades synthetic geometry grew greatly in content on the foundations laid by Poncelet, Steiner, and von Staudt. It was eventually made the subject of a number of standard textbooks, of which Reye's *Geometrie der Lage* (1868, 3rd ed. 1886–92)[24] is one of the best-known examples.

[23]W. Blaschke, *Kreis und Kugel* (Leipzig, 1916), pp. 1–42.
[24]Translated as *Lectures on the Geometry of Position* (New York, 1898).

17

Representatives of algebraic geometry were Möbius and Plücker in Germany, Chasles in France, and Cayley in England. August Ferdinand Möbius, for more than fifty years an observer at, and later director of, the Leipzig Astronomical Observatory, was a scientist of many parts. In his book *Der barycentrischer Calcül* (1827), he was the first to introduce homogeneous coordinates. When the masses m_1, m_2, m_3 are placed at the vertices of a fixed triangle, Möbius gave to the center of gravity (barycentrum) of these masses the coordinates $m_1:m_2:m_3$, and showed how these coordinates are well fitted to describe the projective and affine properties of the plane. Homogeneous coordinates, from this time on, became the accepted tool for the algebraic treatment of projective geometry. Working in quiet isolation not unlike his contemporary von Staudt, Möbius made many other interesting discoveries. An example is the null system in the theory of line congruences, which he introduced in his textbook on statics (1837). The "Möbius strip" (published 1865), a first example of a nonorientable surface, is a reminder of the fact that Möbius is also one of the founders of our modern science of topology.[25]

Julius Plücker, who taught for many years at Bonn, was an experimental physicist as well as a geometer. He made a series of discoveries in crystal magnetism, electrical conduction in gases, and spectroscopy. In a series of papers and books, especially in his *Neue Geometrie des Raumes* (1868–69), he reconstructed analytic geometry by the application of a wealth of new ideas. Plücker showed the power of the abbreviated notation, in which for instance $C_1 + \lambda C_2 = 0$ represents a pencil of conics. In this book he introduced homogeneous coordinates, known as "projective" coordinates based on a fundamental tetrahedron, and also the fundamental principle that geometry need not solely be based on points as basic elements. Lines, planes, circles, spheres can all be used as the elements (*Raumelemente*) on which a geometry can be based. This fertile conception threw new light on both synthetic and algebraic geometry and created new forms of duality. The number of dimensions of a particular form of geometry could now be any positive integer number, depending on the number of parameters necessary to define the "element." Plücker also published a general theory of algebraic curves in the plane, in which he derived the "Plücker relations" between the number of singularities (1834, 1839).

Michel Chasles, for many years the leading representative of geometry in France, was a pupil of the Ecole Polytechnique in the later days of Monge and in 1841 became professor at this institute. In 1846 he accepted the chair of higher geometry, especially established for him, at the Sorbonne, where he taught for many years. Chasles's work had much in common with that of Plücker, notably in his ability to obtain the maximum of geometrical information from his equations. It led him to adroit operation with isotropic lines and circular points at infinity. Chasles followed Poncelet in the use of "enumerative" methods, which in his hands developed into a new branch of geometry, the so-called enumerative geom-

---

[25]The "Möbius strip" was also discovered by J. B. Listing at Göttingen and published by Listing in 1861. Both Möbius and Listing had discovered it in 1858.

etry. This field was later fully explored by Hermann Schubert in his *Kalkül der abzählenden Geometrie* (1879), and by H. G. Zeuthen in his *Abzählende Methoden* (1914). Both books reveal the strength as well as the weakness of this type of algebra in geometrical language. Its initial success provoked a reaction led by E. Study, who stressed that "precision in *geometricis* may not perpetually be treated as incidental."[26]

Chasles had a fine appreciation for the history of mathematics, especially of geometry. His well-known *Aperçu historique sur l'origine et le développement des méthodes en géométrie* (1837) stands at the beginning of the modern history of mathematics. It is a very readable text on Greek and modern geometry, and is also a good example of a history of mathematics written by a productive scientist.[27]

18

During these years of almost feverish productivity in the new projective and algebraic geometries another novel and even more revolutionary type of geometry lay hidden in a few obscure publications discarded by most leading mathematicians. The questions of whether Euclid's parallel postulate is an independent axiom or can be derived from other axioms had puzzled mathematicians for two thousand years. Ptolemy had tried to find an answer in antiquity, Naṣīr al-dīn in the Middle Ages, and Lambert and Legendre in the eighteenth century. All these men had tried to prove the axiom and had failed, even if they reached some very interesting results in the course of their investigations.[28] Gauss was the first man who believed in the independent nature of the parallel postulate, which implied that other geometries, based on another choice of axiom, were logically possible. Gauss never published his thoughts on this subject. The first to challenge openly the authority of two millennia and to construct a non-Euclidean geometry were a Russian, Nikolai Ivanovič Lobačevskiĭ, and a Hungarian, János Bolyai. The first to publish his idea was Lobačevskiĭ, who was a professor in Kazan and lectured on the subject of Euclid's parallel axiom in 1826. His first book appeared in 1829–30 and was written in Russian. Few people took notice of it. Even a later German edition, with the title *Geometrische Untersuchungen zur Theorie der Parallellinien*,[29] received little attention, even though Gauss showed interest. By that time Bolyai had already published his ideas on the subject.

---

[26]See E. Study, *Verhandlungen Dritter Intern. Math. Kongress* (Heidelberg, 1905), pp. 388–95; B. L. van der Waerden, Diss. (Leiden, 1926).

[27]Chasles also received some notoriety as the victim of a document forger, who between 1861 and 1870 managed to sell him thousands of spurious items, from letters written by Pascal to others written by Plato (and even one from Mary Magdalene to Lazarus). See J. A. Farrer, *Literary Forgeries* (London, 1907), Ch. XII.

[28]It is a curious fact that, unacknowledged at the time, the Scotch philosopher Thomas Reid, in his critique of Berkeley's theory of vision, published an outline of a non-Euclidean geometry (of the elliptic type), indicated by light rays: *An Inquiry into the Human Mind* (1764). See N. Daniels, "Thomas Reid's Discovery of Non-Euclidean Geometry," *Philosophy of Science*, Vol. 39 (1972), pp. 219–34.

[29]Translated as "Geometrical Researches on the Theory of Parallels" by G. B. Halsted, and reprinted in R. Bonola, *Non-Euclidean Geometry* (1912), Dover Publications, Inc., 1955.

JAKOB STEINER (1796–1863)

ARTHUR CAYLEY (1821–1895)

NIKOLAI IVANOVIČ LOBAČEVSKIĬ (1793–1856)

János (Johann) Bolyai was the son of a mathematics teacher in a provincial town of Hungary. This teacher, Farkas (Wolfgang) Bolyai, had studied at Göttingen when Gauss was also a student there. The two men kept up an occasional correspondence. Farkas spent much time in trying to prove Euclid's fifth postulate (see Chap. III, Sec. 7, above), but could not come to a definite conclusion. His son inherited his passion and also began to work on a proof despite his father's plea to do something else:

> You should detest it just as much as lewd intercourse, it can deprive you of all your leisure, your health, your rest, and the whole happiness of your life. This abysmal darkness might perhaps devour a thousand towering Newtons, it will never be light on earth . . . (Letter of 1820).

János Bolyai entered the army and built up a reputation as a dashing officer. He began to accept Euclid's postulate as an independent axiom and discovered that it was possible to construct a geometry, based on another axiom, in which through one point in a plane an infinity of lines can be laid which do not intersect a line in the plane. This was the same idea which had already occurred to Gauss and Lobačevskiï. Bolyai wrote down his reflections, which were published in 1832 as an appendix to a book by his father, and which had the title *Appendix scientiam spatii absolute veram exhibens.*[30] The worrying father wrote to Gauss for advice on the unorthodox views of his son. When the answer from Göttingen came, it contained enthusiastic approval of the younger Bolyai's work. Added to this was Gauss's remark that he could not praise Bolyai, since this would mean self-praise, the ideas of the *Appendix* having been his own for many years.

Young János was deeply disappointed by this letter of approval which elevated him to the rank of a great scientist but robbed him of his priority. His disappointment increased when he met with little further recognition, although he continued to write in mathematics, e.g., on a geometrical representation of imaginaries. He was deeply stirred when he became acquainted with Lobačevskiï's book through its German translation of 1840, but continued his retired life till his death in 1860.

Bolyai's and Lobačevskiï's theories were similar in principle, though their papers were very different. It is remarkable how the new ideas sprang up independently in Göttingen, Budapest, and Kazan, and in the same period after an incubation period of two thousand years. It is also remarkable how they matured partly outside the geographical periphery of the world of mathematical research. Sometimes great new ideas are born outside, not inside, the schools. There was a link, however: Gauss, in Göttingen, was a student friend of the older Bolyai, and Lobačevskiï's teacher at Kazan had been J. M. Bartels, who had been one of Gauss's teachers.

Non-Euclidean geometry (the name is due to Gauss) remained for several decades an obscure field of mathematics. Most mathematicians ignored it, the prevailing Kantian philosophy refusing to take it seriously. The first leading scientist to understand its full importance was Riemann, whose general theory of manifolds (1854) made full allowance not only for the existing types of non-Euclidean geometry, but also for many other, so-called Riemannian, geometries. However, full acceptance of these theories came only when the generation after Riemann began to understand the meaning of his theories (1870 and later).

Still another generalization of classical geometry originated in the years before Riemann and did not find full appreciation until after his death. This was the geometry of more than three dimensions. It came fully developed into the world in Grassmann's *Ausdehnungslehre* ("Theory of Extension") of 1844. Hermann Grassmann was a teacher at the gymnasium in Stettin and was a man of extraordinary versatility; he wrote on such varied subjects as electric currents, colors and acoustics, linguistics, botany, and folklore. His Sanskrit dictionary of the *Rigveda* is still in use. The *Ausdehnungslehre*, of which a revised and more readable edition was published in 1861, was written in strictly Euclidean form. It built up a geometry in a space of *n* dimensions, first in affine, then in metrical space. Grassmann used an invariant symbolism, in which we now recognize a vector and

---

[30]Translated as "The Science of Absolute Space" by G. B. Halsted, and reprinted in Bonola, *Non-Euclidean Geometry.*

tensor notation (his "gap" products are tensors), but which made his work almost inaccessible to his contemporaries. A later generation took parts of Grassmann's structure to build up vector analysis for affine and for metrical spaces.

Although Cayley in 1843 introduced the same concept of a space of n dimensions in a far less forbidding form, geometry of more than three dimensions was received with distrust and incredulity. Here again Riemann's address of 1854 made full appreciation easier. Added to Riemann's ideas were those of Plücker, who pointed out that space elements need not be points (1865), so that the geometry of lines in three-space could be considered as a four-dimensional geometry, or, as Klein has stressed, as the geometry of a four-dimensional quadric in a five-dimensional space. Full acceptance of geometries of more than three dimensions occurred only in the latter part of the nineteenth century, mainly because of their use in interpreting the theory of algebraic and differential forms in more than three variables.

19

The names of Hamilton and Cayley show that by 1840 English-speaking mathematicians had at last begun to catch up with their continental colleagues. Until well into the nineteenth century, the Cambridge and Oxford dons regarded any attempt at improvement of the theory of fluxions as an impious revolt against the sacred memory of Newton. The result was that the Newtonian school of England and the Liebnizian school of the continent drifted apart to such an extent that Euler, in his integral calculus (1768), considered a union of both methods of expression as useless. The dilemma was broken in 1812 by a group of young mathematicians at Cambridge who, under the inspiration of the older Robert Woodhouse, formed an Analytical Society to propagate the differential notation. Its leaders were George Peacock, Charles Babbage, and John Herschel. They tried, in Babbage's words, to advocate "the principles of pure d-ism as opposed to the *dot*-age of the university."[31] This movement met initially with severe criticism, which was overcome by such actions as the publication, in 1816, of an English translation of Lacroix's *Traité élémentaire de calcul différentiel et de calcul intégral* (2nd ed., 1806). The new generation in England now began to participate in modern mathematics.

The first important contribution came not from the Cambridge group, however, but from some mathematicians who had taken up continental mathematics independently. The most important of these mathematicians were Hamilton and George Green. It is interesting to notice that with both men, as well as with Nathaniel Bowditch in New England, the inspiration to study "pure d-ism" came from the study of Laplace's *Mécanique céleste*. Green, self-taught, a miller's son from Nottingham, followed the new discoveries in electricity with great care. There was, at that time (c. 1825), almost no mathematical theory to account for electrical phenomena; Poisson, in 1812, had made no more than a beginning. Green read Laplace, and, in his own words:

[31]See J. B. Dubbey, "The Introduction of the Differential Notation in Great Britain," *Annals of Science*, Vol. 19 (1963), pp. 37–48.

Considering how desirable it was that a power of universal agency, like electricity, should, as far as possible, be submitted to calculation, and reflecting on the advantages that arise in the solution of many difficult problems, from dispensing altogether with a particular examination of each of the forces which actuate the various bodies in any system, by confining the attention solely on that peculiar function on whose differentials they all depend, I was induced to try whether it would be possible to discover any general relations, existing between this function and the quantities of electricity in the bodies producing it.

The result was Green's *Essay on the Application of Mathematical Analysis to Theories of Electricity and Magnetism* (1828), the first attempt at a mathematical theory of electromagnetism. It was the beginning of modern mathematical physics in England and, with Gauss's paper of 1839, established the theory of potential as an independent branch of mathematics. Gauss did not know Green's paper, which only became more widely known when William Thomson (the later Lord Kelvin) had it reprinted in Crelle's *Journal* of 1846. Yet the kinship of Gauss and Green was so close that where Green selected the term "potential function," Gauss selected almost the same term, "potential," for the solution of Laplace's equation. Two closely related identities, connecting line and surface integrals, are called the "formula of Green" and the "formula of Gauss." The term "Green's function" in the solution of partial differential equations also honors the miller's son who studied Laplace in his spare time.

We have no room for a sketch of the further development of mathematical physics in England, or, for that matter, in Germany. With this development the names of Stokes, Rayleigh, Kelvin, and Maxwell, of Kirchhoff and Helmholtz, of Gibbs and of many others are connected. These men contributed to the solution of partial differential equations to such an extent that mathematical physics and the theory of linear partial differential equations of the second order sometimes seemed to become identified. Mathematical physics, however, brought fertile ideas to other fields of mathematics, to probability and complex function theory, as well as to geometry. Of particular importance was James Clerk Maxwell's *Treatise on Electricity and Magnetism* (2 vols., 1873),[32] which gave a systematic mathematical exposition of the theory of electromagnetism based on Faraday's experiments. This theory of Maxwell eventually dominated mathematical electricity, and later inspired the theories of Lorentz on the electron and Einstein on relativity.

20

Nineteenth-century pure mathematics in England was primarily algebra, with applications mainly to geometry and with three men—Cayley, Sylvester, and Salmon—leading in this field. Arthur Cayley devoted his early years to the study and practice of law, but in 1863 he accepted the new Sadlerian professorship of mathematics at Cambridge where he taught for thirty years. In the forties, while Cayley was practicing law in London, he met Sylvester, at that time an actuary; and from those years dates Cayley's and Sylvester's common interest in the

---

[32]Reprinted by Dover Publications, Inc., 1954.

algebra of forms—or *quantics*, as Cayley called them. Their collaboration meant the beginning of the theory of algebraic invariants.

This theory had been "in the air" for many years, especially after determinants began to be a subject of study. The early work of Cayley and Sylvester went beyond mere determinants; it was a conscious attempt to give a systematic theory of invariants of algebraic forms, complete with its own symbolism and rules of composition. This was the theory which was later improved by Aronhold and Clebsch in Germany and formed the algebraic counterpart of Poncelet's projective geometry. Cayley's voluminous work covered a large variety of topics in the fields of finite groups, algebraic curves, determinants, and invariants of algebraic forms. Among his best-known works are his nine *Memoirs on Quantics* (1854–78). The sixth paper of this series (1859) contained the projective definition of a metric with respect to a conic section. This discovery led Cayley to the projective definition of the Euclidean metric and in this way enabled him to assign to metrical geometry its position inside the framework of projective geometry. The relation of this projective metric to non-Euclidean geometry escaped the eye of Cayley; it was later discovered by Felix Klein.

James Joseph Sylvester was not only a mathematician but also a poet, a wit, and with Leibniz the greatest creator of new terms in the whole history of mathematics. From 1855 to 1869 he taught at Woolwich Military Academy. He was twice in America, the first time as a professor at the University of Virginia (1841–42), the second time as a professor at Johns Hopkins University in Baltimore (1877–83). During this second period he was one of the first to establish graduate work in mathematics at American universities. With the teaching of Sylvester, mathematics began to flourish in the United States.

Two of Sylvester's many contributions to algebra have become classics: his theory of elementary divisors (1851, rediscovered by Weierstrass in 1868), and his law of inertia of quadratic forms (1852, already known to Jacobi and Riemann, but not published). We also owe to Sylvester many terms now generally accepted, such as invariant, covariant, contravariant, cogredient, and syzygy. Many anecdotes have been told about him—several of the absent-minded-professor variety.

The third English algebraist-geometer was George Salmon, who during his long life was connected with Trinity College, Dublin, Hamilton's alma mater, where he instructed in both mathematics and divinity. His main merit lies in his well-known textbooks, which excel in clarity and charm. These books opened the road to analytical geometry and invariant theory to several generations of students in many countries and even now have hardly been surpassed. They are *Conic Sections* (1848), *Higher Plane Curves* (1852), *Modern Higher Algebra* (1859), and *Analytic Geometry of Three Dimensions* (1862). The study of these books can still be highly recommended to all students of geometry.

21

Two products of the algebra of the United Kingdom deserve our special attention: Hamilton's quaternions and Clifford's biquaternions. Hamilton, the Royal Astrono-

JAMES JOSEPH SYLVESTER (1814–1897)

WILLIAM ROWAN HAMILTON (1805–1865)

mer of Ireland, having completed his work on mechanics and optics, turned in 1835 to algebra. His *Theory of Algebraic Couples* (1835) defined algebra as the science of pure time and constructed a rigorous algebra of complex numbers on the conception of a complex number as a number pair. This was probably independent of Gauss, who in his theory of biquadratic residues (1831) had also constructed a rigorous algebra of complex numbers, but based on the geometry of the complex plane. Both conceptions are now equally accepted. Hamilton subsequently tried to penetrate into the algebra of number triples, number quadruples, etc. The light dawned upon him—as his admirers like to tell—on a certain October day of 1843, when walking by a Dublin bridge he discovered the quaternion. His investigations on quaternions were published in two big books, the *Lectures on Quaternions* (1853) and the posthumous *Elements of Quaternions* (1866). The best-known part of this quaternion calculus was the theory of vectors (the name is due to Hamilton), which was also part of Grassmann's theory of extension. It is mainly because of this fact that the algebraic works of Hamilton and Grassmann are now frequently quoted. In Hamilton's days, however, and long afterwards, the quaternions themselves were the subject of an exaggerated admiration. Some British mathematicians saw in the calculus of quaternions a kind of Leibnizian *arithmetica universalis*, which of course aroused a reaction (Heaviside versus Tait) in which quaternions lost much of their glory. The theory of hypercomplex numbers, elaborated by Peirce, Study, Frobenius, and Cartan, eventually put quaternions in their legitimate place as the simplest associative number system of more than two units. The cult of the quaternion in its heyday even led to an International Association for Promoting the Study of Quaternions and Allied Systems of Mathematics, which disappeared as a victim of World War I. Another aspect of the quaternion controversy was the fight between partisans of Hamilton and Grassmann, when, through the efforts of Gibbs in America and Heaviside in England, vector analysis had emerged as an independent brand of mathematics. This controversy raged between 1890 and World War I, and was finally solved by the application of the theory of groups, which established the merits of each method in its own field of operation.[33]

William Kingdon Clifford, who died in 1879 at the age of thirty-three, taught at Trinity College, Cambridge, and at University College, London. He was one of the first Englishmen who understood Riemann, and with him shared a deep interest in the origin of our space conceptions. Clifford developed a geometry of motion for the study of which he generalized Hamilton's quaternions into the so-called biquaternions (1873–76). These were quaternions with coefficients taken from a system of complex numbers $a + b\epsilon$, where ϵ^2 may be $+1$, -1 or 0, and which could also be used for the study of motion in non-Euclidean spaces. Clifford's *Common Sense of the Exact Sciences* is still good reading; it brings out his kinship in thinking with Felix Klein. This kinship is also revealed in the term "spaces of Clifford-Klein" for certain closed Euclidean manifolds in non-Euclidean

---

[33]F. Klein, *Vorlesungen über die Entwickelung der Mathematik im 19. Jahrhundert* (Berlin, 1927), Vol. II, pp. 27–52; J. A. Schouten, *Grundlagen der Vektor- und Affinoranalysis* (Leipzig, 1914); and the many papers of E. Cartan.

geometry. If Clifford had lived, Riemann's ideas might have influenced British mathematicians a generation earlier than they actually did.

For many decades pure mathematics in the English-speaking countries maintained its strong emphasis on formal algebra. It influenced the work of Benjamin Peirce of Harvard University, a pupil of Nathaniel Bowditch, who did distinguished work in celestial mechanics and in 1870 published his *Linear Associative Algebras*, one of the first systematic studies of hypercomplex numbers.[34] The formalist trend in English mathematics may also account for the appearance of *An Investigation of the Laws of Thought* (1854)[35] by George Boole of Queen's College, Dublin. Here it was shown how the laws of formal logic which had been codified by Aristotle and taught for centuries in the universities could themselves be made the subject of a calculus. It established principles in harmony with Leibniz's idea of a *characteristica generalis*.

Boole was influenced by the work on symbolic logic done by Augustus De Morgan, from 1826 to 1866 professor at University College, London, who also presented a first form of a logic of propositions. De Morgan, in his long career, greatly influenced British mathematics. He was also a wit: see his posthumously published *Budget of Paradoxes* (1872).

The "algebra of logic" opened a school of thought that endeavored to establish a unification of logic and mathematics. It received its impetus from Gottlob Frege's book *Die Grundlagen der Arithmetik* (1884), which offered a derivation of arithmetical concepts from logic. These investigations reached a climax in the twentieth century with the *Principia Mathematica* of Bertrand Russell and Alfred North Whitehead (1910–13); they also influenced the later work of Hilbert on the foundations of arithmetic and the elimination of the paradoxes of the infinite.[36]

22

The papers on the theory of invariants by Cayley and Sylvester received the greatest attention in Germany, where several mathematicians developed the theory into a science based on a complete algorithm. The main figures were Hesse, Aronhold, Clebsch, and Gordan. Hesse, who was a professor at Königsberg and later at Heidelberg and Munich, showed, like Plücker, the power of abbreviated methods in analytical geometry. He liked to reason with the aid of homogeneous coordinates and of determinants. Aronhold, who taught at the Technische Hochschule in Berlin, wrote a paper in 1858 in which he developed a consistent symbolism in invariant theory with the aid of so-called ideal factors (which bear no relation to those of Kummer); this symbolism was further developed by Clebsch in 1861, under whose hands the "Clebsch-Aronhold" symbolism became the al-

---

[34] One of Benjamin Peirce's sons, Charles Sanders Peirce, was not only an able applied mathematician, as was his father (for the U.S. Coast and Geodetic Survey), but made profound contributions to the philosophy of mathematics which have only recently found wider recognition. See, e.g., C. Eiserle, *Studies in the Scientific and Mathematical Philosophy of C. S. Peirce* (The Hague, 1979; a collection of essays).

[35] Reprinted by Dover Publications, Inc., 1951.

[36] D. Hilbert and W. Ackermann, *Grundzüge der theoretischen Logik* (Berlin, 1928). See also M. Black, *The Nature of Mathematics* (New York and London, 1934).

most universally accepted method for the systematic investigation of algebraic invariants. We now recognize in this symbolism, as well as in Hamilton's vectors, Grassmann's gap products, and Gibbs's dyadics, special aspects of tensor, and hence of linear, algebra. This theory of invariants was later enriched by Paul Gordan of the University of Erlangen, who proved (1868–69) that to every binary form belongs a finite system of rational invariants and covariants in which all other rational invariants and covariants can be expressed in rational form. This theorem of Gordan (the *Endlichkeitssatz*) was extended by Hilbert in 1890 to algebraic forms in n variables

Alfred Clebsch was professor at Karlsruhe, Giessen, and Göttingen and died at thirty-nine years of age. His life was a condensation of remarkable achievements. He published a book on elasticity (1862), following the leadership of Lamé and Saint-Venant in France; he applied his theory of invariants to projective geometry. He was one of the first men who understood Riemann and was a founder of that branch of algebraic geometry in which Riemann's theories of functions and of multiply connected surfaces were applied to real algebraic curves. Clebsch-Gordan's *Theorie der Abelschen Funktionen* (1866) gave a broad outline of these ideas. Clebsch also founded the *Mathematische Annalen*, for more than sixty years the leading mathematical journal. His lectures on geometry (1876–77), published by his pupil F. Lindemann, remain a standard text on projective geometry.

23

By 1870 mathematics had grown into an enormous and unwieldy structure, divided into a large number of fields in which only specialists knew the way. Even great mathematicians—Hermite, Weierstrass, Cayley, Beltrami—at most could be proficient in only a few of these many fields. This specialization has constantly grown until at present it has reached alarming proportions. Reaction against it has never stopped, and some of the most important achievements of the last hundred years have been the result of a synthesis of different domains of mathematics.

Such a synthesis had been realized in the eighteenth century by the works of Lagrange and Laplace on mechanics. They remained a basis for very powerful work of varied character. The nineteenth century added to this new unifying principles, notably the theory of groups and Riemann's conception of function and of space. Their meaning can best be understood in the work of Klein, Lie, and Poincaré.

Felix Klein was Plücker's assistant in Bonn during the late sixties; it was there that he learned geometry. He visited Paris in 1870 when he was twenty-two years of age. Here he met Sophus Lie, a Norwegian six years his senior, who had become interested in mathematics only a short time before. The young men met the French mathematicians, among them Camille Jordan of the École Polytechnique, and studied their work. Jordan, in 1870, had just written the *Traité des substitutions*, a book on substitution groups and Galois's theory of equations. Klein and Lie began to understand the central importance of group theory and subsequently divided the field of mathematics more or less into two parts. Klein, as a rule, concentrated on discontinuous, Lie on continuous groups.

In 1872 Klein became professor at Erlangen. In his inaugural address he explained the importance of the group conception for the classification of the different fields of mathematics. The address, which became known as the "Erlangen program," declared every geometry to be the theory of invariants of a particular transformation group. By extending or narrowing the group we can pass from one type of geometry to another. Euclidean geometry is the study of the invariants of the metrical group, projective geometry of those of the projective group. Classification of groups of transformation gives us a classification of geometry; the theory of algebraic and differential invariants of each group gives us the analytical structure of a geometry. Cayley's projective definition of a metric allows us to consider metrical geometry in the frame of projective geometry. "Adjunction" of an invariant conic to a projective geometry in the plane gives us the non-Euclidean geometries. Even relatively unknown topology received its proper place as the theory of invariants of continuous point transformations.

In the previous year Klein had given an important example of his mode of thinking when he showed how non-Euclidean geometries can be conceived as projective geometries with a Cayley metric. This brought full recognition at last to the neglected theories of Bolyai and Lobačevskiĭ. Their logical consistency was now established. If there were logical mistakes in non-Euclidean geometry, then they could be detected in projective geometry, and few mathematicians were willing to admit such a heresy. Later this idea of an "image" of one field of mathematics on another field was often used and played an important part in Hilbert's axiomatics of geometry.

The theory of groups made possible a synthesis of the geometrical and algebraic work of Monge, Poncelet, Gauss, Cayley, Clebsch, Grassmann, and Riemann. Riemann's theory of space, which had offered so many suggestions embodied in the Erlanger program, inspired not only Klein but also Helmholtz and Lie. Helmholtz (in 1868 and 1884) studied Riemann's conception of space, partly by looking for a geometrical image of his theory of colors, and partly by inquiring into the origin of our ocular measure. This led him to investigate the nature of geometrical axioms and especially Riemann's quadratic measurement. Lie improved on Helmholtz's speculations concerning the nature of Riemann's measurement by analyzing the nature of the underlying groups of transformations (1890). This "Lie-Helmholtz space problem" has been of importance not only to relativity and group theory, but also to physiology.[37]

Klein gave an exposition of Riemann's conception of complex functions in his booklet *Über Riemann's Theorie der algebraischen Funktionen* (1882),[38] in which he stressed how physical considerations can influence even the subtlest type of mathematics. In the *Vorlesungen über das Ikosaeder* (1884)[39] he showed that modern algebra could teach many new and surprising things about the ancient Platonic bodies. This work was a study of rotation groups of the regular bodies

---

[37]H. Freudenthal, "Neuere Fassungen des Riemann-Helmholtzschen Raumproblems," *Math. Zeitschrift* 63 (1956), pp. 374–405.

[38]Translated as *On Riemann's Theory of Algebraic Functions and their Integrals*, by Frances Hardcastle.

[39]Translated as *Lectures on the Icosahedron* by G. G. Morrice.

FELIX KLEIN (1849–1925)

MARIUS SOPHUS LIE (1842–1899)

and their relation to Galois groups of algebraic equations. In extensive studies by himself and by scores of pupils, Klein applied the group conception to linear differential equations, to elliptic modular functions, to Abelian and to the new "automorphic" functions, the last in an interesting and friendly competition with Poincaré. Under Klein's inspiring leadership, Göttingen, with its traditions of Gauss, Dirichlet, and Riemann, became a world center of mathematical research where young men and women of many nations gathered to study their special subjects as an integral part of the whole of mathematics. Klein gave inspiring lectures, the notes of which circulated in mimeographed form and provided whole generations of mathematicians with specialized information and—above all—with an understanding of the unity of their science. After Klein's death in 1925 several of these lecture notes were published in book form.

While in Paris Sophus Lie had discovered the contact transformation, and with this the key to the whole of Hamiltonian dynamics as a part of group theory. After his return to Norway he became a professor in Christiania; later, from 1886 to 1898, he taught at Leipzig. He devoted his whole life to the systematic study of continuous transformation groups and their invariants, demonstrating their central importance as a classifying principle in geometry, mechanics, and ordinary and partial differential equations. The result of this work was codified in a number of standard volumes, edited with the aid of Lie's pupils Scheffers and Engel (*Theorie der Transformationsgruppen*, 1888–93; *Differentialgleichungen*, 1891; *Kontinuierliche Gruppen*, 1893; *Berührungstransformation*, 1896). Lie's work was later considerably enriched by, and under the influence of, the French mathematician Elie Cartan.

24

France, faced with the enormous growth of mathematics in Germany, continued to produce excellent mathematicians in all fields. It is interesting to compare German and French mathematicians; Hermite with Weierstrass, Darboux with Klein, Hadamard with Hilbert, Paul Tannery with Moritz Cantor. From the 1840s to the 1860s, the leading mathematician was Joseph Liouville, professor at the Collège de France in Paris, a good teacher and organizer and editor for many years of the *Journal de mathématiques pures et appliquées*. He investigated in a systematic way the arithmetic theory of quadratic forms of two and more variables, but "Liouville's theorem" in statistical mechanics shows him as a productive worker in an entirely different field. He established the existence of transcendental numbers and in 1844 proved that neither e nor e^2 can be a root of a quadratic equation with rational coefficients. This was a step in the chain of arguments which led from Lambert's proof in 1761 that π is irrational to Hermite's proof that e is transcendental (1873) and to the final proof by F. Lindemann that π is transcendental (1882). Liouville and several of his associates developed the differential geometry of curves and surfaces; the formulas of Frenet-Serret (1847) came out of Liouville's circle.

Charles Hermite, a professor at the Sorbonne and at the Ecole Polytechnique, became the leading representative of analysis in France after Cauchy's death in

1857. Hermite's work, as well as that of Liouville, was in the tradition of Gauss and Jacobi; it also showed kinship with that of Riemann and Weierstrass. Elliptic functions, modular functions, Theta-functions, number and invariant theory all received his attention, as the names "Hermitian numbers" and "Hermitian forms" testify. His friendship with the Dutch mathematician Stieltjes, who through Hermite's intervention obtained a chair at Toulouse, was a great encouragement to the discoverer of the Stieltjes integral and the application of continued fractions to the theory of moments. The appreciation was mutual: *Vous avez toujours raison et j'ai toujours tort,*[40] Hermite once wrote to his friend. The four-volume *Correspondance* (1905) between Hermite and Stieltjes contains a wealth of material, mainly on functions of a complex variable.

The French geometrical tradition was gloriously continued in the books and papers of Gaston Darboux. Darboux was a geometer in the sense of Monge, approaching geometrical problems with full mastery of groups and differential equations, and working on problems of mechanics with a lively space intuition. Darboux was professor at the Collège de France and for half a century active in teaching. His most influential work was his standard *Leçons sur la théorie générale des surfaces* (4 vols., 1887–96), which presented the results of a century of research in the differential geometry of curves and surfaces. In Darboux's hands this differential geometry became connected in the most varied ways with ordinary and partial differential equations as well as with mechanics. Darboux, with his administrative and pedagogical skill, his fine geometrical intuition, his mastery of analytical technique, and his understanding of Riemann, occupied a position in France somewhat analogous to that of Klein in Germany.

This second part of the nineteenth century was the period of the great French comprehensive textbooks on analysis and its applications, which often appeared under the name of *Cours d'analyse* and were written by leading mathematicians. The most famous are the *Cours d'analyse* of Camille Jordan (3 vols., 1882–87) and the *Traité d'analyse* of Emile Picard (3 vols., 1891–96), to which was added the *Cours d'analyse mathématique* by Edouard Goursat (2 vols., 1902–05).

25

The greatest French mathematician of the second half of the nineteenth century was Henri Poincaré, who from 1881 until his death in 1912 was professor at the Sorbonne in Paris. No mathematician of his period commanded such a wide range of subjects and was able to enrich them all. Each year he lectured on a different subject; these lectures were edited by students and cover an enormous field: potential theory, light, electricity, conduction of heat, capillarity, electromagnetics, hydrodynamics, celestial mechanics, thermodynamics, probability. Every one of these lectures was brilliant in its own way; together they presented ideas which have borne fruit in the works of others while many still await further elaboration. Poincaré, moreover, wrote a number of popular and semipopular works in which he tried to give a general understanding of the problems of

---

[40]"You are always right and I am always wrong."

T. J. Stieltjes (1856–1894)

Henri Poincaré (1854–1912)

modern mathematics. Among these are *La valeur de la science* (1905),[41] and *La science et l'hypothèse* (1906).[42] Apart from these lectures Poincaré published a large number of papers on the so-called automorphic and Fuchsian functions, on differential equations, on topology, and on the foundations of mathematics, treating with great mastery of technique and full understanding all pertinent fields of pure and applied mathematics. No mathematician of the nineteenth century, with the possible exception of Riemann, has so much to say to the present generation.

The key to the understanding of Poincaré's work may lie in his meditations on celestial mechanics, and in particular on the three-body problem (*Les méthodes nouvelles de mécanique céleste*, 3 vols., 1893). Here he showed direct kinship with Laplace and demonstrated that even at the end of the nineteenth century the ancient mechanical problems concerning the universe had lost nothing of their pertinence to the productive mathematician. It was in connection with these problems that Poincaré studied divergent series and developed his theory of asymptotic expansions, that he worked on integral invariants, the stability of orbits, and the shape of celestial bodies. His fundamental discoveries on the local behavior of the integral curves of differential equations near singularities, as well as their behavior in the large, are related to his work on celestial mechanics. This is also true of his investigations on the nature of probability, another field in which he shared Laplace's interest. Poincaré was like Euler and Gauss: wherever we approach him we discover the stimulation of originality. Our modern theories concerning relativity, cosmogony, probability, and topology are all vitally influenced by Poincaré's work.

26

The Risorgimento, the national rebirth of Italy, also meant the rebirth of Italian mathematics. Several of the founders of modern mathematics in Italy participated in the struggle that liberated their country from Austria and unified it; later they combined political positions with their professional chairs. The influence of Riemann was strong, and through Klein, Clebsch, and Cayley the Italian mathematicians obtained their knowledge of geometry and the theory of invariants. They also became interested in the theory of elasticity with its strong geometrical appeal.

Among the founders of the new Italian school of mathematicians were Brioschi, Cremona, and Betti. In 1852 Francesco Brioschi became professor in Pavia, and in 1862 organized the technical institute at Milan where he taught until his death in 1897. He was a founder of the *Annali di matematica pura ed applicata* (1858)— which indicated in the title its desire to emulate Crelle's and Gergonne's journals. In 1858 in company with Betti and Casorati he visited the leading mathematicians of France and Germany. Volterra later claimed that "the scientific existence of Italy as a nation" dated from this journey.[43] Brioschi was the Italian representative

[41]Translated as *The Value of Science* by G. B. Halsted.
[42]Translated as *Science and Hypothesis* and reprinted by Dover Publications, Inc., 1952.
[43]V. Volterra, *Bull. Am. Math. Soc.*, Vol. 7 (1900), pp. 60–2.

of the Cayley-Clebsch type of research in algebraic invariants. Luigi Cremona, after 1873 director of the engineering school in Rome, has given his name to the birational transformation of plane and space, the "Cremona transformations" (1863–65). He was also one of the originators of graphical statics.

Eugenio Beltrami was a pupil of Brioschi and occupied chairs in Bologna, Pisa, Pavia, and Rome. His main work in geometry was done between 1860 and 1870 when his differential parameters introduced a calculus of differential invariants into surface theory. Another contribution of that period was his study of so-called pseudospherical surfaces, which are surfaces of which the Gaussian curvature is negative constant. On such a pseudosphere we can realize a two-dimensional non-Euclidean geometry of Bolyai. This was, with Klein's projective interpretation, a method to show that there were no internal contradictions in non-Euclidean geometry, since such contradictions would also appear in ordinary surface theory.

By 1870 Riemann's ideas became more and more the common property of the younger generation of mathematicians. His theory of quadratic differential forms was made the subject of two papers by the German mathematicians E. B. Christoffel and R. Lipschitz (1870). The first paper introduced the "Christoffel symbols." These investigations, combined with Beltrami's theory of differential parameters, brought Gregorio Ricci-Curbastro in Padua to his so-called absolute differential calculus (1884). This was a new invariant symbolism originally constructed to deal with the transformation theory of partial differential equations, but it provided at the same time a symbolism fitted for the transformation theory of quadratic differential forms.

In the hands of Ricci and of some of his pupils, notably Tullio Levi-Civita, the absolute differential calculus developed into what we now call the theory of tensors. Tensors were able to provide a unification of many invariant symbolisms, and also showed their power in dealing with general theorems in elasticity, hydrodynamics, and relativity. The name tensor has its origin in elasticity (W. Voigt, 1900).

The most brilliant representative of differential geometry in Italy was Luigi Bianchi. His *Lezioni di geometria differenziale* (2nd ed., 3 vols., 1902–09) ranks with Darboux's *Théorie générale des surfaces* as a classical exposition of nineteenth-century differential geometry.

27

David Hilbert, professor at Göttingen, presented to the International Congress of Mathematicians in Paris in 1900 a series of twenty-three research projects. At that time Hilbert had already received recognition for his work on algebraic forms and had prepared his now famous book on the foundations of geometry— *Grundlagen der Geometrie*, 1900. This book was in many respects inspired by the pioneering work of Moritz Pasch of Giessen, especially by his book *Vorlesungen über neuere Geometrie* (1882), in which Pasch extended to the foundations of geometry the axiomatic mode of thinking which, at the same time, led Frege to his work on the foundations of arithmetic. Hilbert, in his book, gave an analysis of the axioms on which Euclidean geometry is based and explained how modern axiomatic research has been able to improve on the achievements of the Greeks.

In this address of 1900 Hilbert tried to grasp the trend of mathematical research of the past decades and to sketch the outline of future productive work.[44] A summary of his projects will give us a better understanding of the meaning of nineteenth-century mathematics. As it also looks forward to developments in twentieth-century mathematics, I have placed this summary in the next chapter (Sec. 2).[45]

Hilbert's program demonstrated the vitality of mathematics at the end of the nineteenth century and contrasts sharply with the pessimistic outlook existing toward the end of the eighteenth century. At present some of Hilbert's problems have been solved; others still await their final solution. The development of mathematics in the years after 1900 has not disappointed the expectations raised at the close of the nineteenth century. Even Hilbert's genius, however, could not foresee some of the striking developments that actually have taken place and are taking place today. Twentieth-century mathematics has followed its own novel path to glory.

Literature

The best history of nineteenth-century mathematics is:

Klein, F. *Vorlesungen über die Entwickelung der Mathematik im 19. Jahrhundert I, II.* Berlin, 1926–27. (English translation of Part I with appendices, Brookline, Mass., 1979.)

A bibliography of leading nineteenth-century mathematicians is given in:

Sarton, G. *The Study of the History of Mathematics.* Cambridge, Mass., 1936, pp. 70–98. (Lists biographies and editions of collected works for all the chief mathematicians of the nineteenth and twentieth centuries. Further bibliographical material is given in the issues of *Scripta mathematica*, 1932–present.)

Furthermore, see:

de Launay, L. *Monge. Fondateur de l'École Polytechnique.* Paris, 1934.
Taton, R. *Monge.* Paris, 1951. (Shortened version in *Elemente der Mathematik*, Suppl. 49, Basel, 1950.)
Dunnington, G. W. *Carl Friedrich Gauss: Titan of Science.* New York, 1956.
Gauss, C. F. *Gedenkband anlässlich des 100. Todestages,* ed. H. Reichardt. Leipzig, 1957.
Klein, F. *Materialen für eine wissenschaftliche Biographie von Gauss.* 8 vols., Leipzig, 1911–20.

---

[44]Translation in *Bull. Am. Math. Soc.*, 2nd ser., Vol. 8 (1901–02), pp. 437–79.

[45]A discussion of the problems outlined by Hilbert after thirty years appears in L. Bieberbach, "Über den Einfluss von Hilberts Pariser Vortrag über 'Mathematische Probleme' auf die Entwicklung der Mathematik in den letzten dreissig Jahren," *Naturwissenschaften*, Vol. 18 (1936), pp. 1101–11. More recent: *Die Hilbertschen Probleme,* ed. by P. S. Alexandrov, Ostwalds Klassiker, Vol. 252 (Leipzig, 1971; from the Russian).

Worbs, E. *Carl Friedrich Gauss: Ein Lebensbild.* Leipzig, 1955.

Quaternion Centenary Celebration. *Proc. Roy. Irish Acad. A*, Vol. 50 (1945), pp. 69–98. (Among the articles is "The Dublin Mathematical School in the First Half of the Nineteenth Century," by A. J. McConnell.)

"A Collection of Papers in Memory of Sir William Rowan Hamilton." *Scripta mathematica Studies*, New York, 1951.

Kötter, E. "Die Entwicklung der synthetischen Geometrie von Monge bis auf von Staudt." *Jahresber. Deut. Math. Verein*, Vol. 5 (1901), pp. 1–486.

Black, M. *The Nature of Mathematics*. New York, 1934. (Contains a bibliography on symbolic logic.)

Kagan, V. F. *Lobačevskiĭ*. Moscow and Leningrad, 1945 (in Russian). (French translation, Moscow, 1974.) (See I. Toth, *HM*, Vol. 6 (1979), pp. 91–97.)

One Hundred Five and Twenty Years of Non-Euclidean Geometry of Lobačevskiĭ. A. P. Norden, ed. Moscow and Leningrad, 1952 (in Russian.)

Merz, J. T. *A History of European Thought in the Nineteenth Century.* 4 vols., London, 1903–14.

Hadamard, J. *The Psychology of Invention in the Mathematical Field*. Princeton, 1945. Dover reprint, 1954.

Prasad, G. *Some Great Mathematicians of the Nineteenth Century: Their Lives and Their Works.* 2 vols., Benares, 1933–34.

Struik, D. J. "Outline of a History of Differential Geometry." *Isis*, Vol. 19 (1933), pp. 92–120; *ibid.*, Vol. 20 (1934), pp. 161–91.

Coolidge, J. L. "Six Female Mathematicians." *Scripta math.*, Vol. 17 (1951), pp. 20–31. (Concerning Hypatia, M. G. Agnesi, E. du Châtelet, M. Somerville, S. Germain, and S. Kovalevsky.)

Kovalevsky, Sonja. *Her Recollections of Childhood*. New York, 1895. (Trans. from the Russian by I. F. Hapgood. Contains also the biography by A. C. Leffler, from the Swedish [1892], which exists in other translations, e.g., one in German in the Reclam ed., Leipzig.)

In Remembrance of S. V. Kovalevskaya. A Collection of Essays. Moscow, 1951 (in Russian). (See also *Istor.-mat. Issled.*, Vol. 7 [1954], pp. 666–715.)

Wheeler, L. P. *Josiah Willard Gibbs*. New Haven, 1951.

Kollros, L. "Jakob Steiner." In *Elemente der Mathematik*, Suppl. 7, Basel, 1947.

Winter, E. *B. Bolzano und sein Kreis*. Leipzig, 1933; Halle, 1949.

Kolman, E. *Bernard Bolzano*. Berlin, 1963.

Ore, O. *Niels Henrik Abel. Mathematician Extraordinary.* Minneapolis, 1957.

Infeld, L. *Whom the Gods Love.* New York, 1948. (A novel about Galois.)

Dalmas, A. *Évariste Galois révolutionnaire et géomètre.* Paris, 1956.

Biermann, K. R. "J. P. G. Lejeune Dirichlet, Dokumente für sein Leben und Wirken." *Abh. Deutsch. Akad. Wiss., Klass für Math.*, No. 2 (1959), pp. 1–68.

——. "Der Mathematiker Ferdinand Minding und die Berliner Akademie." *Monatsberichte Deutsch. Akad. Wiss.*, 3 (1961), pp. 120–33.

Medvedev, F. A. *The Development of the Theory of Sets in the 19th Century.* Moscow, 1965 (in Russian).

——. *The Development of the Concept of Integral.* Moscow, 1974 (in Russian). (See *HM*, Vol. 6 [1979], pp. 85–90.)

Manning, K. "The Emergence of the Weierstrassian Approach to Complex Analysis." *AHES*, Vol. 14 (1975), pp. 297–383.

[Kolmogorov, A. N., and Juškevič, A. P., eds.] *Mathematics of the 19th Century: Mathematical Logic, Algebra, Theory of Numbers, Theory of Probability.* Moscow, 1978 (in Russian).

Scholz, E. *Geschichte des Mannigfaltigkeitsbegriffs von Riemann bis Poincaré.* Boston, etc., 1980.

Biermann, K. R. *Gotthold Eisenstein.* Crelle 214/251 (1964), p. 1920.

Dugac, P. *Richard Dedekind et les fondements des mathématiques.* Paris, 1976.

——. "Eléments d'analyse de Karl Weierstrass." *AHES*, Vol. 10 (1973), pp. 41–176. (Bibliography, pp. 297–383.)

I. Grattan-Guinness. *The Development of the Foundation of Mathematics from Euler to Riemann.* Cambridge, Mass., 1970.

Herivel, J. *Joseph Fourier, the Man and the Physicist.* Oxford, 1975. (Cf. I. Grattan-Guinness, *Annals of Science*, Vol. 32 [1975], pp. 503–14.)

Dauben, J. W. *Georg Cantor, His Mathematics and Philosophy of the Infinite.* Cambridge, Mass., 1979. (See also P. E. B. Jourdain, *Arch. Math. Phys.*, Vol. 3, pp. 10, 14, 16, 22.)

Rozenfel'd, B. A. *History of Non-Euclidean Geometry.* Moscow, 1976 (in Russian). (See *HM*, Vol. 6 [1979], pp. 460–64.)

Morrison, P. and E. *Babbage's Calculating Machine or Differential Engine.* New York, 1965.

Grabiner, J. V. *The Origins of Cauchy's Rigorous Calculus.* Cambridge, Mass., and London, 1981.

[Rüdenberg, L., and Zassenhaus, H., eds.] *Hermann Minkowski: Briefe an David Hilbert.* Berlin, etc., 1973. (The letters from Hilbert to Minkowski have not [yet?] been recovered.)

Reid, C. *Hilbert.* New York, 1970.

Métivier, M., Costabel, P., and Dugac, P. *Siméon-Denis Poisson et la science de son temps.* Paris, 1981.

Marx, K. *Matematičeskie Rukopisi.* Moscow, 1968. (Karl Marx's mathematical manuscripts, German with Russian translation and commentary.)

Kennedy, H. C. "Karl Marx and the Foundations of the Differential Calculus." *HM*, Vol. 4 (1977), pp. 303–18. (See also *Science and Nature*, Vol. 1 [1978], pp. 59–62.)

Bos, H. J. M., and Mehrtens, H. "The Interactions of Mathematics and Society in History. Some Explanatory Remarks." *HM*, Vol. 4 (1977), pp. 7–30, with extensive bibliography.

CHAPTER IX

The First Half of the Twentieth Century

1

When the twentieth century opened, mathematics was in flourishing condition, although creative mathematics was still in the main confined to one section of the world, was in most cases an academic profession, and was restricted, with few exceptions, to white males of European stock. The leading countries remained France and Germany. In France the center was Paris; in Germany, less centralized, it was Göttingen, with other universities such as Berlin running pretty close. Distinguished mathematical work also came out of Russia, Great Britain, Italy, Switzerland, Scandinavia, Belgium, and the Netherlands, while the U.S.A. and Japan began to show that Europe, though leading, no longer possessed the monopoly it had maintained since the Renaissance. Few persons would disagree that the leading mathematicians were Klein and Hilbert in Göttingen, and Poincaré in Paris, but influential also were Volterra in Italy, Darboux and Hadamard in France, and Minkowski in Zurich (and soon also in Göttingen), to mention a few.

Academies remained active, some, like the Académie des Sciences in Paris, very active. Most mathematicians, however, earned their income in educational institutions, those engaged in research to be found on university faculties. Some, as in Scandinavia and the Netherlands, were advisers to insurance companies, but, although polytechnical institutes and technical universities had mathematical staffs for training engineers, thus providing a close relationship with industry, few mathematicians were directly engaged in production. Such employment was only beginning to occur. Charles P. Steinmetz, who studied at Breslau and Zurich, and from 1895 on was consulting engineer at General Electric in Schenectady, New York, applied complex function theory to alternating currents, as did Arthur Kennelly, who from 1902 on taught engineering at Harvard (and later also at MIT). From the 1880s on, Oliver Heaviside, in England, had already been applying the calculus to electromagnetism in industry, where he became known for his operational analysis and the "telegraph equation." He remained a private citizen, his seaside cottage a "hermitage." Kennelly's and Heaviside's names are linked in the name of the layer of atmosphere also known as the ionosphere.

Felix Klein, conscious of the growing importance of mathematics to industry, was instrumental in obtaining organizational and financial support from private interests for research in applied mathematics for physicists and engineers. One result was the Institute for Aerodynamical and Hydrodynamical Research at Göttingen, with Ludwig Prandtl, a mechanical engineer, in charge (1908). Such institutes were still rare at the time.

For the mathematicians of this period, we must therefore look to the universities. Like their fellow professionals, mathematicians had organized or were organizing special societies. Two survived from earlier days, those of Hamburg (1690) and Amsterdam (1776). Among the newer mathematical societies were those of Moscow (1860), London (1865), Paris (1872), Edinburgh (1883), Palermo (1884), Berlin (1899), New York (1888, growing in 1894 into the American Mathematical Society). Others were to follow: one in India in 1907 and another in 1908, and one in Spain in 1911; organization in Poland began in 1911. Mathematicians could thus meet each other eye-to-eye at regular intervals.

The first international gathering of some importance took place in 1893 during the Columbian Exposition in Chicago, where Klein presented a series of lectures. The next meeting, known as the First International Congress, followed in 1897 at Zurich. There were about 200 participants, the congress languages being German and French. One of the main addresses was by Adolf Hurwitz of Zurich, dealing with analytic functions and Cantor's set theory, still a relative novelty. Two equally modern topics entered the discussion: the logical foundations (Schröder, Peano) and the functions of functions (Volterra), for which Hadamard suggested the term "fonctionnelles."

The next congress, again on the occasion of a universal exposition, was that of Paris in 1900, and has remained in the collective memory of the profession because of Hilbert's twenty-three problems. There were many professional meetings at Paris in that year; one of them was the First International Congress of Philosophy, attended by Peano, Whitehead, and Russell, who discussed the logical foundations of mathematics. Mathematics and philosophy, having parted ways during the nineteenth century (with some exceptions, as with Riemann and Boole) were finding each other again. As Emile Picard had said in 1897 at Zurich, "Les mathématiques sont en grande coquetterie avec la philosophie." But which philosophy?

The next international congresses were at Heidelberg (1904), Rome (1908), and Cambridge, England (1912). Because of the First World War and the tensions it raised, the first truly international congress did not convene until 1928, at Bologna.

The necessity of an overview of the rapidly advancing and widely different fields of pure and applied mathematics led Klein and some of his German colleagues to that great enterprise, the *Encyklopädie der mathematischen Wissenschaften*. It started publication in 1898 and continued to be published until 1935 as a collection of monographs, endeavoring, with some success, in the spirit of Klein, to develop the interrelationship of the different domains of mathematics. Its focus ranged from that of Section I, *Arithmetik und Algebra*, to that of Section VI, 2, *Astronomie*. In 1904 a revised edition began to appear in French, but this became a victim of the First World War. Those who liked a shorter collection of monographs could use the *Repertorio* (1897-1900), edited by Ernesto Pascal, professor at Pavia and later at Naples. This served as prototype of the German *Repertorium der höheren Mathematik* (5 vols., 1910-29), with writings by several authors on geometry and analysis.

Another form of mathematics reporting appeared in the *Jahrbuch über die Fortschritte der Mathematik*. This was first published in 1871 with short reports on publications that had appeared up to 1868, and was continued year by year. For the year 1900 it listed 2,000 items from about 1,500 authors. The time interval between the reports in the *Jahrbuch* and the publications themselves being three years, the Amsterdam Mathematical Society began to publish the *Revue semestrielle des publications mathématiques* in 1892. This seldom reported more than titles, but the time interval was much shortened. It lasted till 1938.

The number of mathematical journals had increased since the appearance of the Crelle and Liouville journals. Among them we notice the *Annali di matematica* (1858), the *Matematičeskiǐ Sbornik* (Moscow, 1866), the *Mathematische Annalen* (1868), the *Bulletin des sciences mathématiques* (1870), the *American Journal of Mathematics* (1878), the *Acta Mathematica* (Sweden, 1882), the *Rendiconti di Palermo* (1885), and the *Transactions of the American Mathematical Society* (1899). From a later date are the *Mathematische Zeitschrift* (1918) and the Polish *Fundamenta Mathematica* (1920). All these journals have continued to exist. Academies also offered their own publications, as well as some schools such as the Paris Ecole Normale and later also MIT. To keep up with all this information required considerable effort, especially since Latin as the international *lingua franca* had disappeared with Gauss and Jacobi. Yet a paper published in such a prestigious journal as the *Mathematische Annalen* could find a wide audience.

Many books published in those days are now dated, but some have retained their appeal. We think, for example, of books by Hilbert, Hausdorff, Borel, Russell and Whitehead, Lebesgue, and Sierpiński.

2

The Dutch historian Jan Romein, in a richly documented study,[1] has drawn our attention to the many structural changes that occurred between roughly 1890 and 1910 in almost all fields of human endeavor, from economics to history and music, all in the wake of the social transformations that would lead to the catastrophe of 1914. Mathematics was no exception, mainly, in its case, as a result of its inner dynamics. The main factors in this transformation, themselves interdependent, may be seen in the growing penetration of Cantor's theory of sets (aggregates) into many fields of mathematics—not without difficulties, even protests[2]—and in the closely related investigations into the foundations of mathematics and the

---

[1]Cf. J. Romein, *The Watershed of Two Eras, Europe in 1900* (Middletown, Conn., 1978). Translation from the Dutch (Leiden and Amsterdam, 1967).

[2]"Aus dem Paradies, das Cantor uns geschaffen, soll uns niemand vertreiben können" (Hilbert, *Math. Annalen*, Vol. 95 [1926], pp. 161-90; cf. Chap. VIII, Sec. 14 of this book): "From the paradise Cantor has created for us, nobody should be able to expel us."

development of abstract structures in algebra, logic, and general spaces. More and more the ancient concept of mathematics as the theory of quantity was abandoned; more and more it was seen as the theory of structure in general. Among new fields opened were the Lebesgue theory of integration, functional analysis, operational calculus, and tensors, and there were debates between formalists, intuitionists, and logisticians. But the development was not purely intrinsic; a mighty impact came from mathematical physics, where after 1905 quantum theory and relativity challenged the wits of mathematicians, physicists, astronomers, philosophers, and even chemists and theologians. There were also other outside sources of change, such as biology (biometrics) and engineering.

The leading figure of the older generation was David Hilbert in Göttingen, especially after the death of Poincaré in 1912 and the increasing concentration on fields of education by his colleague, Felix Klein, in the decades before the latter's death in 1925. A good idea of the status of mathematics in the period around 1900 is obtained from a study of Hilbert's famous Paris problems of 1900, already alluded to in the previous chapter of this book; and so we shall give here a summary of these powerful delineations of areas for further research.

1. *Cantor's problem of the cardinality of the continuum.* Is the continuum the cardinality next to the denumerable set—and can the continuum be considered as well ordered?

2. *The consistency of (lack of contradictions among) the arithmetical axioms.* If this consistency exists, then the consistency of geometrical axioms can be established.

3. *The equality of two tetrahedra in volume, if base, area, and altitude are equal.* Prove this with the aid of division and combination alone (hence without infinitesimals).

4. *The problem of the straight line as the shortest connection of two points.* This question is raised, for example, by the geometry of Minkowski and by certain problems in the calculus of variations.

5. *Lie's concept of the continuous transformation group without postulating the differentiability of the functions defining the group.* This question may lead to functional equations.

6. *Mathematical treatment of the axioms of physics.* From the axioms of geometry we can pass to those of rational mechanics (as, e.g., Boltzmann did in 1897) and to such fields as statistical mechanics, probability, etc.

7. *Irrationality and transcendence of certain numbers.* Examples are numbers of the form α^β for algebraic $\alpha \neq 0$ and an algebraic irrational β, such as $2^{\sqrt{2}}$ or $e^\pi = i^{-2i}$—are they transcendental or irrational? Hilbert thought of Hermite and Lindemann on π (see Chapter VIII, Sec. 24).

8. *Problems in prime number theory.* Here we think of Riemann's Zeta function and Goldbach's conjecture that every even number is at least in one way the sum of two primes (1742, letter to Euler).

9. *Proof of the most general law of reciprocity in arbitrary number fields.* This referred immediately to some recent work by Hilbert on relative quadratic number fields.

10. *Decision whether a Diophantine equation with integral rational numbers i. solvable in such numbers.* This was an ancient problem already tackled for some equations of higher degree than the second and related to Fermat's "great problem."

11. *The theory of quadratic forms with algebraic coefficients.* This again had a direct bearing on Hilbert's work on number fields.

12. *Generalization of Kronecker's theorem on Abelian fields to an arbitrary field of rationality.* This brings us to a domain where algebraic functions, number theory, and abstract algebra meet.

13. *The impossibility of solving the general equation of degree seven by means of functions of only two variables.* This was a problem suggested by nomography, as Maurice d'Ocagne had explained it.[3]

14. *The proof of the finite character of certain systems of "relative integral" functions.* Extending here the notion of integral functions to *relativganz*, this problem asks for the generalization of the theorems on the finiteness in the classical theory of invariants due to Hilbert and Gordan.

15. *Rigorous foundation of Schubert's enumerative geometry.* For this a firm foundation in algebra will be necessary.

16. *The problem of the topology of algebraic curves and surfaces.* Solution of this problem is only in its infancy, though we have some knowledge, especially in the case of curves.

17. *The representation of definite functions* (functions never negative for real value of the variables) *by quotients of sums of squares of functions.*

18. *Construction* (filling) *of space by congruent polyhedra.* This relates to a question of group theory and crystallography, and the work of E. S. von Fedorov and A. Schoenfliesz.[4]

19. *Are the solutions of regular variational problems always analytic?* Here "regular" is specifically defined. Hilbert remarks that every surface of positive constant curvature has to be analytic, but this does not hold for surfaces of negative constant curvature.

20. *The general boundary problem,* especially demonstrating the existence of solutions of partial differential equations with given boundary values, and generalizations of regular variational problems.

21. *Proof of the existence of linear differential equations with prescribed monodromy group.* This problem was suggested by Poincaré's theory of Fuchsian functions.

22. *Uniformization of analytic relations by means of automorphic functions.* This

---

[3]D'Ocagne, of the Paris École Polytechnique, is considered the creator of this science of solving equations by graphic tables (he also created the name). His *Nomographie* appeared in 1891, followed in 1899 by his *Traité de nomographie*. The principle is much older; we think of the work of Junius Massau of Ghent (1884).

[4]E. S. von Fedorov was a mine director in the Urals. Arthur Schoenfliesz, later a professor at Frankfurt, was with Klein in Göttingen when the works of Fedorov and his own work (the 230 crystallographic groups in space) appeared, independently, in 1890 and 1891. See *AHES*, Vol. 4 (1967), pp. 235–40. Fedorov, in 1891, also discovered that there were just seventeen two-dimensional symmetry groups of repeating patterns (as on wallpaper). Rediscovered by G. Polya and P. Niggli in 1924; see H. S. M. Coxeter, *Introduction to Geometry* (New York, 1981), Chap. 4.

was also suggested by Hilbert's proof that the uniformization of any algebraic relation between two variables can be accomplished by means of automorphic functions of one variable.

23. *Extension of the methods of the calculus of variations.* Hilbert added this "propaganda" suggestion because he found that despite Weierstrass' contributions in this field it still contained many angles that had been poorly investigated and were potentially useful in several fields of mathematics and mechanics (such as the three-body problem).

"Now you really have taken a total lease on mathematics for the twentieth century" ("*du hast die Mathematik für das 20.te Jahrhundert in Generalpacht genommen*"), wrote Hilbert's friend Minkowski to him from Zurich after the delivery of the Paris address. Though this remark may sound somewhat melodramatic, the truth is that the subjects indicated in these twenty-three problems have persistently stimulated research in depth up to the present day. Some of the problems have been solved: number 3 by Max Dehn (in 1904; he showed that the proof is not always possible), number 17 by Emil Artin (1920). Some have been solved in part, such as number 7 (e.g., by A. Gelfond, 1924); this is understandable, since these "problems" are more in the nature of programs. Such is number 16, indicating the possibility of a whole new domain of mathematics. And the wide use to which the calculus of variations has been put, not only in pure mathematics, but in such fields as relativity, shows that a good reason existed for the inclusion of Problem 23.[5]

The year 1900 also marked the publication of the two-volume report for the German Mathematical Society by Schoenfliesz on the development of the theory of point sets. It dealt with the applications to the theory of functions of a real variable and with the theory of integration, as well as with the concept of the measure of point sets, where the author discussed several approaches, including those of Cantor, Peano, Jordan, and Borel. It was in connection with this concept that further progress was made, now in France.

3

In the later part of the nineteenth century the theory of real functions had made fundamental progress, especially in such concepts as functional dependence, integration, and differentiation, often in connection with the investigation of trigonometric series. This investigation had led to new conditions for integration as well as to Cantor's theory of sets. Names connected with these studies are Paul DuBois Reymond in Berlin, Ulisse Dini in Pisa, Camille Jordan in Paris.

Jordan, in the 1880s and later, introduced, especially in his *Cours d'analyse* (3 vols., 1882–84),[6] the concept of functions of bounded variation, and brought in, as Poincaré did about the same time, topological considerations. He asked for a rigorous proof of "Jordan's theorem" that a simple closed curve in the plane

---

[5]See Chap. VIII, footnote no. 45, above; also the summary by H. Freudenthal, *DSB*, Vol. VI (1972), pp. 393–94. A general solution of this ancient and challenging problem was given by K. F Sundman in Helsinki between 1907 and 1912, but it is not practical for numerical computation (now performed by computers).

[6]Third ed., 1909–15.

divides it into an outside and an inside part. He placed integration within the context of a "measurable" set.

This idea was further developed by Emile Borel at the Paris Ecole Normale, where as a student around 1890 he was "extrêmement séduit" by Cantor's theories. In 1894, in his thesis, he introduced the "Heine-Borel" covering theorem[7] and the proof that a denumerable set is of measure zero, measure here defined from a finite set of intervals to a more extensive set ("Borel measure"). In 1898 he published his *Leçons sur la théorie des fonctions*.[8]

Here Henri Lebesgue's work began, also at the Ecole Normale and also in the 1890s, when he was a student. Later, after having occupied some provincial teaching positions, he returned to Paris in 1910, first at the Sorbonne, then (1921) at the Collège de France.

Lebesgue, at the Ecole Normale, was in close contact with the four-years-older Borel and his contemporary René Baire.[9] Baire's thesis of 1899, *Sur les fonctions de variables réelles*, was able to bring Cantor's theory of sets to bear on limit functions of continuous functions, introducing functions of different "Baire class." His thesis was followed by that of Lebesgue of 1902: *Intégrale, longueur, aire*. Directly appealing to the work of Jordan and Borel, Lebesgue introduced his notion of measure: "no more fundamental subject than this," he wrote in 1931, referring to the role this concept has played in man's history. Basing his work upon his now standard notion of measure, Lebesgue defined his integration, which definition is now also standard, since it brought unity to this field. One of his theorems was that a continuous function of bounded variation possesses a finite derivative except possibly on a set of measure zero.

Lebesgue's further work, taken up by other mathematicians—such as Maurice Fréchet, also of the Ecole Normale—made his theories, originally received by many mathematicians with a certain skepticism (Why bother with all these unusual, "pathological," functions?) more and more acceptable. The Lebesgue integral removed many difficulties encountered since the time that Riemann had defined his integral and Weierstrass his theory of real functions. It was Fréchet who introduced abstract spaces (1908), and Baire's pupil Arnaud Denjoy, a generalization of the Lebesgue integral. Fréchet's ideas were again taken up by others, among them Stefan Banach, developing in and after 1920 the "Banach spaces."

It is said that young Lebesgue's lack of satisfaction with the function theory of his day is illustrated by his remark that a crumpled handkerchief must be a ruled surface since his Nancy professor had taught that a surface is applicable on a plane if and only if it is developable and therefore composed of straight lines. Among other such "anomalies" was also the "Peano curve" (1890), a mapping of a line segment by continuous functions x and y of a variable t on a square, hence a curve

---

[7]The combination of these names has been thought unwise. Eduard Heine, a professor at Halle, had in 1872 announced a theorem equivalent to that of Borel, but only Borel pointed out its significance.

[8]New ed., 1950.

[9]Lebesgue and Baire, with Gauss and Monge, belong to the rather few outstanding mathematicians of the past with a working-class (artisan) background. Luzin's grandfather was a serf. Newton had a farming background. Most leading mathematicians of the seventeenth, eighteenth, and nineteenth centuries came out of the middle or upper-middle class.

filling a plane figure. Several other "anomalous" curves were discovered by, e.g., Hilbert. Such discoveries led to the question of the definition of a curve, and from there to that of dimension.

An important contribution to the further penetration of point-set topology into the main body of mathematics was the *Grundzüge der Mengenlehre*, published in 1914 by Felix Hausdorff, a teacher at Bonn.[10] It offered an axiomatic definition of what became known as topological space.

Another generalization of the space concept, so typical of these years, became known as the Hilbert space. It is defined with a metrical measure.

We said that many mathematicians of an older generation were skeptical about the interest taken in "pathological" functions so different from the "smooth" functions they were accustomed to. *"Je me détourne avec effroi et horreur de cette plaie lamentable des fonctions qui n'ont pas de dérivées,"* Hermite once wrote to Stieltjes.[11] They considered the new trend as a kind of mathematical teratology. But the "anomalies" continued to present themselves, such as, for example, the Peano curve (1890).

Giuseppe Peano, from 1890 to his death in 1932 professor at Turin, was a pioneer in symbolic logic and the axiomatic method, stressing the necessity of rigor. This led him to his *Formulario matematico* (5 vols., 1895–1908), a comprehensive presentation of the theorems of mathematics (they came to 4,200), kept logically precise with the aid of his syllogisms. He did not quite convince the mathematical world, but his influence on questions of logic and mathematics has been unmistakable.

4

The theory of real functions was making significant advances. We have mentioned the impact made by investigations concerning trigonometric series, a subject to which Lebesgue applied his theory in a book of 1906. Another domain was functional analysis and especially the theory of integral equations. The name "fonctionnelle" dates, as we have seen, from Jacques Hadamard (1897); it replaced the more narrow term, "line functions." This idea—to study more explicitly functions depending on functions (such as curves), rather than on numbers (or points) as variables—came especially from Italy, where it was Vito Volterra, pupil of Betti and Dini at Pisa, professor at Turin from 1893 to 1900 and then for forty years at Rome, who introduced his "functions of lines" in 1889. He was led, as in many other aspects of his work, by physical considerations, such as the fact that the energy of a current is dependent on the shape of the wire that is moving or bending in an electric field.

Hadamard, and later Fréchet, took their starting point in the calculus of variations, and this was one way in which Fréchet came to his abstract spaces that

---

[10]Rewritten as *Mengenlehre* (1927). English translation: *Set Theory* (1957, from the 3rd ed. of 1937).

[11]"I turn away with fright and horror from this lamentable plague of functions that have no derivatives." I myself heard nothing about Lebesgue during my stay at Leiden University around 1916, and even in 1925, at Göttingen, some of the faculty showed little appreciation. Meeting Wiener there, I first became aware of the importance of Lebesgue's work.

have elements even more abstract than functions. Hadamard, one of the most influential mathematicians of his day—from his doctoral thesis of 1892 on the analytic continuation of a Taylor series, to well into the 1950s (he lived almost ninety-eight years)—was active in many different domains, from logic and number theory to hydrodynamics. His *Série de Taylor et son prolongement analytique* of 1901 has been called the "Bible" of those fascinated by the subject and was indeed long an influential book in this field.

One of the ways by which functional analysis came into being was the study of integral equations. This subject was an old one; we think of Laplace's transform (1792), Abel's integral equation (1823), and those of Liouville (1832 and later). But it was particularly the theory of boundary values in potential theory and other fields in which differential equations were central (such as oscillations in a continuum) that led up to systematic investigation. Volterra introduced the linear type of integral equation that carries his name in 1887. Later, at Stockholm in 1900 and 1903, Ivar Fredholm introduced the same type of equation. They both gave the solution to this type of equation, but it was Fredholm rather than Volterra who made the impact, and this mainly because one of his pupils gave a lecture on his work in Hilbert's seminar during the winter of 1900–01. The analogy of a linear integral equation with a set of n linear equations in n variables was particularly appealing.[12]

Hilbert's mind caught fire at once. He saw the connection with potential theory and the construction of Green's functions for given boundaries, and saw the ascertainment of *eigenvalues* and *eigenfunctions* (these curious bastard terms show the influence of the German contribution) as a problem analogous to the reduction of quadratic forms in n variables to canonical form, as well as the relation to series of orthogonal functions. This again could lead to infinite matrices, all of which concepts played and would continue to play a role in mathematical physics. Thus Hilbert's *Grundzüge einer allgemeinen Theorie der Integralgleichungen* of 1912 added a new domain to mathematics. Many older results, enriched by newer ones, could be seen in a broader context. Abstract spaces spanned by vectors of finite length led to abstract metrical spaces, named after Hilbert. Another example was the Riesz-Fischer theorem on convergence in the mean (1907), named after the Hungarian F. Riesz and E. Fischer of Cologne. Riesz contributed greatly to functional analysis, combining Borel-Lebesgue ideas with the German ones, as is seen in his later book *Leçons d'analyse fonctionnelle* (1952).[13] We also think of Banach's contributions.

This period also brought new results in the classical domain of real and complex analytical functions, obtained with such success by Poincaré and Picard, and continued by Borel, Hadamard, and others. Hadamard applied his results to the analytical theory of numbers, and, studying Riemann's Zeta function, proved that $\pi(x)$, the number of prime numbers being $\leqslant x$, is asymptotically equal to $x/\log x$,

---

[12]Hermann Weyl has made the remark that "Fredholm's discovery has always seemed to me one that was long overdue when it came. What could be more natural than the idea that a set of linear equations connected with a discrete set of mass points gives way to an integral equation when one passes to the limit of a continuum?" (*Amer. Math. Monthly*, Vol. 58 [1951].) Fredholm's equation can be written $\phi\,(x)\,+\,\int_0^1 f(x,y)\,\phi\,(y)dy\,=\,\psi(x)$, ϕ being the unknown.

[13]Written with B. Szökafnalvy-Nagy, translated into English as *Functional Analysis* (1955).

thus verifying a suggestion of Gauss. The same year (1896) brought also a different proof of this "prime number theorem" by the Louvain professor Charles de la Vallée Poussin. (Both men were of the same age, thirty, and lived almost equally long.) Poussin later sharpened the theorem and proved Legendre's suggestion that the log x in the formula should be log $x - 1.08366$. His *Cours d'analyse infinitésimale* (2 vols., first published in 1903–06) was long a standard work, with many editions.

Borel, at the Sorbonne from 1909 to 1940, wrote several monographs on analytical functions, such as the one in 1917 on monogenous functions—those having a derivative at every point of their domain. From 1898 to 1952 he edited the *Collection de monographies sur la théorie des fonctions*, in over fifty volumes, ten by Borel himself, as well as other *Collections*, including one in seven volumes on probability (1937–50).

Poincaré, in 1883, had established that an analytical relation between two variables can be uniformized, that is, the variables can be expressed by one valued "automorphic" function of one variable, but a satisfactory proof of this general uniformization theorem (see Hilbert's twenty-second problem) was only given in 1908 by Poincaré himself and by Paul Koebe, a Göttingen graduate, later professor at Jena and Leipzig, who also reached results in conformal mapping.

Nine years younger than Poincaré, Paul Painlevé, professor first at Lille (1887), then at Paris (1892), turned, after a thesis on analytic functions, to algebraic curves and differential equations and their singular points, the results of which he applied to rational mechanics. In him we meet a mathematician of a type not uncommon in France and Italy: a person prominent at the same time in politics and in science. Painlevé became minister of education in 1915 and of war in 1917, and as prime minister in the same year had Foch appointed representative to the Supreme Allied Council. He was also a pioneer aviator and taught aeronautics, and was again prime minister in 1925. Among his works is a two-volume *Leçons sur la résistance des fluides non visqueux* (1930–31).

5

The trend toward generalization and deeper abstraction typical of so much of the mathematics of this period also extended to the U.S.A., where Eliakim Hastings Moore,[14] from 1892 a professor at the newly founded University of Chicago, was one of the first to create an American school of mathematics—as Čebyšev had done for the Russia of the 1870s and Sierpiński would do for Poland in the decade after 1910. Like so many other American students of his generation, Moore went to Germany to study; at Berlin he was impressed by the rigor of Kronecker and Weierstrass. He worked in many fields, from axiomatics to integral equations, but his name is most prominently connected with his "general analysis," a development of a theory of classes of functions on a general range, influenced by Cantor

---

[14]Not to be confused with his pupil, Robert Lee Moore, after 1920 at the University of Texas, distinguished through his contributions to axiomatics and topology; or with Clarence L. E. Moore, from 1904 to 1931 at MIT and active in differential geometry and tensor calculus.

and Russell, and general enough to underlie and unify particular theories, supported by an adequate notation.[15]

Moore was a successful teacher and administrator. Among his pupils were Robert L. Moore, Oswald Veblen, George D. Birkhoff, and Leonard E. Dickson, representatives of the first generation of leading U.S. mathematicians trained in their own country.

R. L. Moore and Veblen were American representatives of that field of mathematics first known as analysis situs and after 1900 more and more as that part of topology called *combinatory* (in contrast with point-set topology) as it was explained, for instance, by Hausdorff. Its emergence from a set of puzzle-like problems,[16] such as that of Euler on the seven bridges of Königsberg, or the Möbius strip, came with the Riemann surfaces in complex function theory, with Jordan's closed curve theorem, and in particular with Poincaré's publications, between 1895 and 1904, on simplexes, complexes, and Betti numbers of manifolds.

This theory of homology with its group considerations, chains, and cycles was taken up by many investigators, among them Veblen and his colleague at Princeton, James W. Alexander. Of special importance here were the discoveries of the young Dutchman L. E. J. Brouwer, who made his debut with his Amsterdam thesis *On the Foundations of Mathematics* (1907, in Dutch),[17] but then, under the influence of Hilbert's problems (especially the fifth), turned toward several fields, including continuous groups and topology. Between 1908 and 1912 he found his fixed-point theorems with the proof that any continuous mapping of an n-dimensional sphere into itself leaves at least one fixed point invariant. Ever since Cantor and the Peano curve, the question of the invariance of dimension had been coming up. Brouwer now proved that two manifolds of different dimensions cannot be homeomorphic—that is, no one-to-one continuous mapping is possible. This was Brouwer's "invariance theorem" (1910). He also showed the possibility of dividing a circular disc into three regions with the same contour. Much of what is known about the topology of this period can be learned from Veblen's *Analysis situs* (1922), and from *L'analysis situs et la géométrie algébrique* (1924) of Veblen's colleague (after 1928) S. Lefschetz.

6

In this period algebra changed its ancient character. Instead of merely encompassing the theory of algebraic equations and the associated theory of invariants and covariants, it became the abstract doctrine of today with its rings, fields,

[15]For this purpose he urged Florian Cajori, Swiss-born American historian of mathematics, to write his two-volume *History of Mathematical Notations* (1928–29).

[16]Puzzles as mathematical recreations have been popular for many centuries, not rarely yielding their objects in due time to "serious" mathematics. Well-known books with puzzles of this kind are F. E. A. Lucas, *Récréations mathématiques* (4 vols., Paris, 1891–94; reprinted Paris, 1960); W. Ahrens, *Mathematische Unterhaltungen und Spiele* (Leipzig, 1901); and H. E. Dudeney, *Amusements in Mathematics* (London, 1917; Dover reprint, 1958). This tradition was more recently nobly pursued by Martin Gardner in *Scientific American*.

[17]English translation by A. Heyting, 1975.

ideals, and related concepts. One of the origins of the newer algebra was the development of group theory from the Galois theory of algebraic equations into an abstract theory in its own right, especially the theory of finite groups, thus setting a model for the transformation of algebra as a whole. We can study these evolutions in the *Lehrbuch der Algebra* (2 vols., 1895–96) by Heinrich Weber, a professor at Königsberg and later at Strasbourg (at Königsberg he had Hilbert and Minkowski as students).[18] Weber's book has special chapters on groups and algebraic fields. Frege and Peano also did their pioneering work in this subject, the consequences of which were examined by Ernst Steinitz (who was then at Breslau) in his *Algebraische Theorie der Körper* (1910). In this book the field (*Körper*) was the central abstract concept, a system of elements that had two operations, addition and multiplication, and that satisfied associative, commutative, and distributive laws. Steinitz's program was to investigate all such possible fields. Steinitz also mentioned an influence on this work: Kurt Hensel's *Theorie der algebraischen Zahlen* (1908; Hensel taught at Marburg) with its study of the field of "*p*-adic" numbers.

From Steinitz the development of this new algebra went on, especially in the period between the wars, under the influence of Emmy Noether, the daughter of the Erlangen professor Max Noether. Max was known for "Noether's theorem" on algebraic curves (1873) and was a colleague of Emmy's thesis sponsor Paul Gordan (1907). In 1915 Emmy began to lecture at Göttingen under the auspices of Hilbert, but as a woman and a Jew had to struggle against great prejudices. She lost her poorly paid *Lehrauftrag* in algebra with the advent of Hitler and taught from 1933 to her death in 1935 at Bryn Mawr College near Philadelphia. At Göttingen, with her pupils, she developed a general theory of commutative rings and a theory of ideals (inspired by Dedekind) and modules over rings, as well as the fundamental problems of noncommutative algebras, all in strict axiomatic fashion.[19] Among her pupils were Emil Artin, Richard Brauer, and Bartel R. L. van der Waerden, whose widely used *Moderne Algebra* (1930) was inspired by lectures of Artin (in Hamburg) and Emmy Noether. In the USSR, the new algebra was stimulated by Otto Schmidt, also known as a geophysicist and organizer of polar research.

There exist many connections between these developments in algebra and other domains, especially algebraic geometry and the theory of sets. Steinitz himself pointed out that certain theorems could not be demonstrated with the *Auswahlprinzip*, the principle of choice with which the name of Zermelo is connected.

Ernst Zermelo, then at Göttingen, published his well-ordering theorem in 1902. This theorem states that in every set a relation $a < b$ ("a comes before b") can be

---

[18]Weber, with his friend Dedekind, was the editor of Riemann's works (1876) and was also the editor of Riemann's lectures on partial differential equations (1900–01), a book long famous as "Riemann-Weber." It lost some of its glory only after the publication, in 1924 and later, of the still-famous "Courant-Hilbert": *Methoden der mathematischen Physik*. With his Strasbourg colleague J. Wellstein and others Weber also published a three-volume *Encyklopädie der Elementar-Mathematik* (1903–07). There also appeared an Italian parallel in the *Enciclopedia delle mathematiche elementari*, edited by L. Berzolari of Pavia (1930–50).

[19]*Abstrakter Aufbau der Idealtheorie in algebraischen Zahl- und Funktionenkörpern* (*Math. Ann.*, Vol. 96 [1927]).

introduced such that for any two elements a and b either $a = b$, $a < b$, or $b < a$, and that for three elements a, b, c, if $a < b$ and $b < c$, then $a < c$, while every subset has a first element. This theorem came as a result of the general influence of Cantor's theory, a theory in which Cantor himself had left many open questions. One of these was the question of dimension (Hilbert's first problem); another one, this possibility of well-ordering. Since Zermelo based his proof on the *Auswahlaxiom*, which states that from each subset of a given set one of its elements can be singled out, mathematicians differed on the acceptability of a proof where no constructive procedure could be given for the finding of such an element. Hilbert and Hadamard were willing to accept it; Poincaré and Borel were not.

This controversy was only part of the general discussion on the nature of proof brought about by Cantor's set theory, which led to contradictions, to certain "paradoxes"—contradictions in the very foundations of mathematics! Something similar had appeared twice before in history: the Pythagorean discovery of the irrational that was inconsistent with the nature of number (*arithmos*); and the contradictions in the foundation of the calculus at the time of Newton, where a quantity h or dx had to be zero and not zero in the same operation. Eventually, the contradiction was removed in a dialectical way, by means of a compassing theory. In both cases, most mathematicians did not care much, if at all, and went merrily on, convinced that, after all, their science was "true." Would the same thing happen again?

The paradoxes resulting from Cantor's theory were of a different nature, but one example will suffice: that of Bertrand Russell (1903). Let S be the set of all sets that are not members of themselves. Question: is S a member of itself? If so, then it is not a member of itself; if not so, then it is a member of itself. This reminds us of the ancient paradox of the Cretan who said that all Cretans lie. It became clear that you had to handle the set theory with caution, especially when using the term "all," and avoid being semantically careless.

Several attempts were made to indicate the truth value of mathematics. One way was to establish a set of axioms for Cantor's set theory. This was Zermelo's achievement in 1908. His system of seven axioms used only two technical terms, "set" and "$\in$" ("element of"), and added a restricting formulation on the definiteness of the property of a subset, this to avoid the Russell paradox. Axiom 6 was the axiom of choice. Zermelo left aside the question of independence and consistency. Adolf Fraenkel, then at Marburg, and Thoralf Albert Skolem, at Oslo, improved upon this, but Axiom 6 remained a point of discussion, especially after the critique by Kurt Gödel (1930 and later).[20]

Fraenkel became known, even outside the charmed circle of professionals, by his elegant *Einleitung in die Mengenlehre* (1919; extended ed., 1923), a book that originated during the First World War, while he was in the trenches with his fellow

---

[20]Zermelo's work affected many fields. I remember hearing Alfred Pringsheim, in his day a famous after-dinner speaker (like J. L. Coolidge in the U.S.A.), declare in one of his *Bierreden* that function theory *ist bekleidet mit dem Zermelin der Mengenlehre* ("is clothed with the zerm[el]ine of set theory"; *Ermelin* [German] = ermine, stoat [English]). Incidentally, Pringsheim, a function theoretician of the Weierstrass school, and from 1901 till the Nazi days professor at Munich, was the father-in-law of Thomas Mann.

soldiers—a situation reminiscent of Poncelet and his comrades in a Russian prison camp after 1812. After 1929, Fraenkel taught in what was to become Israel.

Hilbert, who in his book on the foundations of geometry (1899) had reduced the consistency of the axioms of geometry to those of arithmetic, was deeply interested in the consistency of the axioms of arithmetic, now under a cloud in the debate on constructivity and apparent contradictions in its foundations. For this purpose he conceived a method called *formalism:* the reduction of mathematics to a finite game with an infinite, finitely defined, apparatus of formulas. The rules of the game must be consistent; the play can never reach a contradiction of the kind $0 = 1$. This led to a field of thought called *metamathematics*, or theory of proof, a science (or philosophy) on a level where formalized mathematics can be studied, avoid vicious circles, and eliminate inconsistencies.

Hilbert's ideas, later laid down in a book written with W. Ackermann (1928), and in another with Paul Bernays (1934),[21] met with considerable skepticism. The sharpest criticism came from L. E. J. Brouwer, who entered the arena in 1907 with his thesis and claimed that truth through constructivity rather than consistency was the essence of mathematics. Between 1913 and 1919, Brouwer developed his *intuitionism*, in which mathematics is seen as starting with the *Ur-Intuition*, a basic intuition of the natural numbers. Only such entities are allowed of which a method of construction can be given. In this process it is not necessary to accept the notion of the excluded middle for infinite sets. This intuitionism, rejecting a fair amount of classical mathematics, gave way to an often acrimonious debate in the 1920s, in which Hermann Weyl, then at Zurich (he had written his thesis under Hilbert), took the side of Brouwer. Weyl at that time had already done distinguished work in integral equations and boundary-value problems, and in his book *Die Idee der Riemannschen Fläche* (1913), aided by Brouwer's topological theorems, sharpened his definitions in complex-function theory. Weyl later somewhat modified his position with respect to foundation questions; his ideas can be studied in his book *Philosophy of Mathematics and Natural Science* (1949), based on an article he wrote in 1926.[22]

Although most mathematicians refused to follow Brouwer in rejecting those parts of mathematics that did not fit in with his demand for constructiveness, they could not but agree on the superiority of an actual constructive procedure to a mere definition, even one consistent with the axioms. After Hilbert's program was shown as unrealizable by Kurt Gödel in 1931, Brouwer's intuitionism was able to survive in renewed form, especially through the work of his fellow Dutchman Arend Heyting (1930 and after).

The Gödel paper that gave a crushing blow to Hilbert's hope, "Über formal unentscheidbare Sätze der Principia Mathematica und verwandter Systeme,"[23]

[21]*Grundzüge der theoretischen Logik* (1928); *Grundlagen der Mathematik* (1934).

[22]H. Weyl, "Philosophie der Mathematik und Naturwissenschaft," in R. Oldenburg, *Handbuch der Philosophie* (1926).

[23]("On formally undecidable propositions of the Principia Mathematica and related systems") *Monatshefte für Mathematik und Physik*, Vol. 38 (1931), pp. 173–98 (translated into English in 1962). See E. Nagel and J. R. Newman, *Gödel's Proof* (New York, 1958).

appeared before Hilbert's *Grundlagen* of 1934. Its principal result was that, in the case that an arithmetical system *S* is without contradictions, this freedom of contradictions cannot be proved within the means of this system. This paper, dealing with completeness, decision, and consistency, opened a new period of work in foundation questions.

The *Principia Mathematica* (3 vols., 1910–13) was written by Bertrand Russell and Alfred North Whitehead, both at Cambridge, under the influence of Frege, Cantor, and Peano. It was the culmination of a program called *logistics* that differed from Hilbert's formalism in its attempt to construct the whole of mathematics by logical deduction from a small number of concepts and principles. The volumes were written in a complicated but precise symbolism, and those who have studied it have admired its logical beauty. Like Hilbert's approach, however, it failed in its ultimate purpose, though its contribution to mathematical logic (like that of Hilbert) has been considerable.

Whitehead, eleven years older than Russell (who was born in 1872 and died in 1970), had already written a *Universal Algebra* (1898) based on Grassmann, Boole, and Hamilton, and had also written on the axioms of projective and descriptive geometry (1906, 1907). In 1924, he settled at Harvard University and became widely known as a philosopher. His *Science and the Modern World* (1925) has a strong Platonic flavor. Russell had also written an *Essay on the Foundations of Geometry* (1897).

8

The foundations of geometry, the subject of work by Pasch, Whitehead, and Russell, became the subject of general interest in the mathematical world and even far beyond that world with the appearance of Hilbert's *Grundlagen der Geometrie*. First published in 1899, it passed through many editions, and was revised and amended after Hilbert's death in 1943 by his longtime collaborator Paul Bernays at Zurich (9th ed., 1962), himself a profound *Grundlagenforscher*. The book opened the road to fresh investigations into the foundations not only of Euclidean, but also of other geometries. An example is the Göttingen theorem of Max Dehn (1903) on the Archimedean postulate as necessary for Legendre's theorem (that the sum of the angles of a plane triangle is not greater than two right angles); later Dehn solved Hilbert's third Paris problem.

Hilbert refers in the preface of his book to the Padua professor Giuseppe Veronese. Veronese was one of the first to construct a non-Archimedean geometry and was also a pioneer in the metrical and projective theory of hyperspaces S_n, where a surface in five-space carries his name (its projection in three-space is the Steiner surface; see Chap VIII, Sec. 16, above). Veronese's colleague Corrado Segre investigated linear transformations and algebraic surfaces (especially quadrics) in S_n; one of his pupils was Julian Lowell Coolidge, after 1908 at Harvard, and also a student at Bonn with Eduard Study. Study was known as a critic of loose formulations in geometry and, with Segre, one of the first to take a rigorous view of the geometry of the complex domain.

We also meet a Dutchman in this Italo-German domain of S_n: Pieter Hendrik

Schoute of Groningen, expert on regular polytopes such as the tessaract (hypercube of S_4). Several of his papers were written in collaboration with Alice Boole Stott, a daughter of George Boole, the logician.[24]

An entirely different and more fundamental approach to hyperspace goes back to Riemann's paper of 1854 with his introduction of a topological manifold, endowing it with a quadratic linear element as for which $ds^2 = g_{ij}\,dx^i dx^j$ (to use our present notation). This led, through the works of Christoffel, Lipschitz, and Beltrami, to the absolute differential calculus of Gregorio Ricci-Curbastro of Padua (1883 and later). A summary of his research, with its wide applications to differential invariants, differential geometry of S_n, elasticity, and rational mechanics, was laid down in a paper in the *Mathematische Annalen*, Vol. 54 (1901), entitled "Méthodes de calcul différentiel absolu," written together with his pupil Tullio Levi-Civita, who contributed his own research. This paper became quite famous after 1913, when Einstein adopted that calculus for his general relativity and gave this *calcul absolu* the name of *tensor calculus*. Einstein's adoption led to a great number of investigations in which tensors were applied to many problems, not only in relativity, and the notion of tensor calculus was widened beyond that of Riemannian geometry. This was greatly spurred by the introduction by Levi-Civita, in a paper of 1917, of the parallelism named after him. Here new ideas were brought in by Weyl (1918) and A. S. Eddington (1923), and the whole new calculus was systematized by J. A. Schouten in Delft (who discovered the parallelism independently in 1918) in his *Ricci-Calcül* (1924, in German, entirely rewritten in an English edition of 1954).

Starting from the point of view of Lie group theory, the subject of his Paris thesis of 1894, Elie Cartan, after 1909 at Paris, was led through his study of differential systems to his method of exterior differential forms.[25] To this he attached, in an entirely original way, the differential geometry of general manifolds, introducing also, after 1923, topological methods on global properties of Lie groups. His papers and books on "spaces of Euclidean, affine, and projective connection" show a remarkable mastery of different domains, combining, in the tradition of Monge and Darboux, analysis with geometry, and the converse.[26]

The term *tensor* was introduced, in its modern meaning, by Woldemar Voigt, Göttingen physicist and crystallographer, in 1908. This brings us to the so-called vector analysis (partaking of both algebra and calculus). Starting from ideas in the

---

[24]George Boole had five daughters, all highly talented. One of them, Mary Ellen, married the Princeton mathematics professor C. H. Hinton, author of a semipopular book, *The Fourth Dimension* (1909). Another daughter, Ethel Boole Voynich, became known, especially in Russia, as the author of the novel *The Gadfly* (1897).

Some new theorems were added in those days to elementary geometry, among them the theorem of Frank Morley, for many years widely discussed (and proved), that the three points of intersection of the adjacent trisectors of the angles of a triangle form an equilateral triangle (c. 1899, and first propagated orally). See H. S. M. Coxeter's *Introduction to Geometry* (1967), pp. 23–25. Morley, British-born, became professor at Johns Hopkins in Baltimore.

[25]See E. Goursat, *Leçons sur le problème de Pfaff* (1922).

[26]E.g., see E. Cartan, *Leçons sur les invariants intégraux* (1922); *Leçons sur la théorie des espaces à connexion projective* (1937). His correspondence with Einstein on manifolds with "teleparallelism" (or "absolute parallelism"), between 1929 and 1932, was published in 1979 by Princeton University Press.

works of W. R. Hamilton and H. Grassmann in the 1830s and 40s, vector analysis was developed in the 1880s by J. W. Gibbs and O. Heaviside as a tool for engineers and physicists. By the end of the century, this idea caught on and texts on it were produced with regularity, beginning in Germany with a book on Maxwell theory by A. Föppl in Leipzig (1894). Gibbs's ideas were expanded by his pupil E. B. Wilson in *Vector Analysis* (1901). Many authors in this field, including Gibbs, extended the concept of vector to other direct quantities, and enriched the field with such notions as polar and axial vectors, dyadics, affinors, rotors, tensors, etc. With the notation differing from country to country, the result was a considerable confusion, increased by the introduction of new notions when, after 1905, special relativity extended these concepts to four dimensions. A theoretical foundation for this calculus was usually missing. Clarification began when Klein, around 1908 (followed by J. A. Schouten in 1914), suggested classification of direct quantities on the basis of group theory, thus reaching a position that Cartan, in his quiet, independent way, had also reached. With general relativity after 1912 came the possibility of using Ricci's tensor calculus as the overall method to understand this field; moreover, its foundations could also be clarified. [27]

9

This period also witnessed a further expansion of the classical theory of numbers. In analytical number theory, we have already mentioned the results of Hadamard and de la Vallée Poussin on the distribution of prime numbers. At Göttingen the representative in this field was Edmund Landau, known for his terse, Euclid-like formulations, for example those in his *Handbuch der Lehre von der Verteilung der Primzahlen* (1909). In England we meet G. H. Hardy and J. E. Littlewood, about whom we shall speak later; and in Russia, G. F. Voronoï. Voronoï contributed to the geometry of numbers. This had been introduced as a new field by Hermann Minkowski in his *Geometrie der Zahlen* (1896; 2nd ed., 1910), a result of his work in ternary quadratic forms which led to theories of convex bodies and the "packing" of spheres (and other bodies).

Minkowski (recipient of the Grand Prix of the Paris Academy of Science at the age of eighteen for the representation of an integer as a sum of five squares of integers [1881]) was from 1896 to 1902 professor at Zurich, and then joined his friend Hilbert on the Göttingen faculty, where he taught till his death in 1909 at the age of forty-five. Deeply interested in electrodynamics, he surprised the learned world with his Cologne lecture *Raum und Zeit* (1908), where he placed Einstein's special relativity into a four-dimensional "Minkowski space" with the bold words: "From this hour on, space and time by themselves will sink completely into shadows and only a kind of union of the two will retain autonomy." [28] This "proclamation" opened the way, not only to Einstein's general relativity, but also to a manifold of research in pure mathematics inspired by it.

---

[27]See R. Weitzenböck, *Invariantentheorie* (1923), for the relations with the classical theory of invariants and covariants; O. Veblen and J. H. C. Whitehead, *The Foundations of Differential Geometry* (1932), began to clarify the topological foundations.

[28]"Von Stund an sollen Raum für sich und Zeit für sich völlig zu Schatten herabsinken und nur noch eine Art Union der beiden soll Selbständigkeit bewahren."

10

The World War of 1914–18 interrupted and often destroyed international relations. Mathematics was no exception. German mathematicians denounced the French (though Hilbert refused to do so); their French colleagues reciprocated. It was *Kultur* versus *culture*. Some mathematicians, such as Volterra, advised their governments. Veblen led a team of mathematicians at the ballistics proving ground at Aberdeen (Maryland), but only during the Second World War did mathematics reach a prominent position in the military, which has since become even stronger. *The International Commission on the Teaching of Mathematics,*[29] founded at the Rome congress of 1908, with Klein presiding, petered out after a promising conference in Paris in April 1914.[30] We have already mentioned the fate of the French *Encyclopédie.*

The first international congress after the war was held at Strasbourg in 1920 (a provocative choice; the city had just been repossessed by France) and excluded the defeated nations. The same thing happened at Toronto in 1924. But at the Bologna congress of 1928, with S. Pincherle of Bologna presiding, the discrimination was abandoned. Europe was still dominant; among the 836 delegates there were fifty-two from outside, and they came from the U.S.A. The International Mathematical Union, founded in 1919 at Brussels, here also became truly international. The Soviet Union participated with thirty-seven delegates in 1928. At the Zurich congress in 1932 there were 667 participants from forty countries, including sixty-six from the U.S.A., and ten from the USSR. The next international congress was in 1936 at Oslo, smaller than before (487 participants, twenty-seven countries); with Hitler the world crisis was intensified. Then came the Second World War, and the next international congress was convened only in 1950.

Mathematics might still be centered in the traditional parts of Europe, but both the U.S.A. and the newly formed USSR were making rapid advances. Harvard, Princeton, Chicago, Moscow, and Leningrad were already important centers. A strong school of mathematics, concentrating on topology and foundation questions, existed in Poland, and vigorous research continued in Hungary and Italy, till in those places (especially Poland), as in Germany and Austria, fascism began to cast its shadow. Now modern mathematics was also to be seen coming from Canada, Japan, Australasia, and India. Total mathematical publication rose steadily.

Specialization revealed itself in the publication of journals devoted to particular fields. The *Fundamenta Mathematica*, founded 1920 in Poland, was devoted to topology and foundations; the German *ZAMM* (*Zeitschrift für Angewandte Mathematik und Mechanik*) appeared in 1921. Series of specialized monographs appeared: the *Mémorial des sciences mathématiques* in France, the *Ergebnisse der exakten Wissenschaften* and the *Grundlehren* (Springer's "yellow series") in Ger-

[29]Often called IMUK in Germany. It was revived at the Bologna conference of 1928.
[30]There were 160 delegates from seventeen countries. Among the participants were Castelnuovo (presiding), Borel, Darboux, D'Ocagne, and Stäckel. See the *Compte-Rendu*, by H. Fehr (Geneva), of 1914; and the follow-up report, by R. C. Archibald (Providence, R.I.), of 1918, with reports from eighteen countries.

many, the *Monografie Matematyczne* in Poland. There were international conferences on special subjects, such as the conference on applied mathematics and mechanics in Delft (1924), and one on tensors (1934) and one on topology (1935) in Moscow.

11

Göttingen, during the years of the Weimar Republic, kept the leading role it had played ever since Klein's arrival in 1886. Here Hilbert was still enthroned as the grand old man of mathematics, even after his official retirement in 1930. The faculty was strong, with Edmund Landau in number theory; Gustav Herglotz in several fields of analysis, such as differential and integral equations; Richard Courant, Klein's successor, applying Klein's and Hilbert's ideas to boundary problems after Dirichlet's principle; and, after 1931, Hermann Weyl, who came from Zurich as Hilbert's successor and whose vision, like that of Hilbert (his thesis adviser), spanned well-nigh the whole of mathematics. From 1902 to his early death in 1909, Minkowski was also on the faculty. At Göttingen were also Paul Bernays, collaborator with Hilbert in his attempt to rebuild the foundations of mathematics, and Emmy Noether, pioneer in the new algebra. The applied aspects were represented by Ludwig Prandtl (aerodynamics) and Carl Runge (numerical problems), as well as by Felix Bernstein, of the Cantor-Bernstein equivalence theorem in set theory, after 1921 head of the Institute of Mathematical Statistics. Nearby was an equally outstanding physics department, where Max Born led a group of young physicists into the era of quantum mechanics—two of these young physicists being Werner Heisenberg and Wolfgang Pauli. Students and visitors continued to flock to this Mecca, filling its classes and its seminars. One of the books emerging from this hive of activity was the already mentioned "Hilbert-Courant": *Methoden der mathematischen Physik* (1924 and later), which appeared just at the moment that the new physics was demanding the type of mathematics (such as boundary problems) presented in it.

Hilbert, sixty years old in 1922, had been called from Königsberg to Göttingen (at Klein's initiative) in 1896. His early work was in algebraic invariants and algebraic number theory, where his *Zahlbericht*, presented to the German Mathematical Society in 1897, became for decades *the* report. Then, around 1900, came the foundations of geometry, around 1906 the calculus of variations, Dirichlet's principle, and the establishment of integral equations as a special field. Then, after an axiomatic interest in mathematical physics (c. 1913, in relativity), came finally, after 1918, his work in the foundations of mathematics (the formalist school). He not only established important theoretical fields, but also engaged in the solution of such special problems as Waring's theorem—the theorem that in three-space every surface of constant negative curvature has singularities. His formalist approach to the foundations should not let us forget his active interest in mathematical physics. Through his publications, his many students, and his famous seminar, he influenced the whole mathematical world and gave direction both to its subjects and to its strict, logical way of thinking. He died in 1943. As already mentioned, Hilbert's successor after his retirement in 1930 was Hermann Weyl.

Weyl had been Hilbert's student, and from 1913 to 1930 taught at Zurich. He followed Hilbert in studies on integral equations and boundary problems, connected with Sturm-Liouville problems and operators in Hilbert space. In 1913 appeared his *Idee der Riemannschen Fläche.* He then turned to the general theory of relativity with his influential *Raum-Zeit-Materie* (1918), after which he explored many fields, especially group theory and its relation to quantum physics. He expressed his ideas in a number of books and papers, continuing his activity after 1933 at Princeton's Institute for Advanced Study. His already quoted *Mathematics and Natural Sciences* (1949) gives a detailed view of his understanding of mathematics and the natural world, with a historic-philosophical background.

We have already mentioned Emmy Noether as a pioneer in the new algebra of ideals. She was—and is—the most outstanding woman mathematician in the history of mathematics. She started at Erlangen, working in Gordan's invariant theory of algebraic (here: ternary biquadratic) forms (1907); then at Göttingen, where she arrived in 1915 at Hilbert's invitation, she first joined him and Klein in their axiomatic-group approach to the new Einstein theory. But her main thrust was in the new abstract algebra. With all the prejudices against her, she never got a higher title than *nichtbeamteter ausserordentlicher Professor.*[31] In 1928–29, she taught in Moscow. Her power to understand the heart of a problem in an abstract way was exceptional.

Strong mathematics departments also existed in other German universities. At Berlin we find I. Schur in algebra and group theory, and Erhardt Schmidt, of the "Schmidt orthogonalization procedure" in Hilbert space (1907). On the Munich faculty, after 1924, was Constantin Carathéodory, Berlin-born of Greek descent, known for his studies in the calculus of variations and complex mapping and uniformization; he edited Euler's *Methodus inveniendi* for the *Opera omnia.*

Much of this flowering came to an end after the Nazis came to power in 1933. In particular, Göttingen suffered, with many of its best members gone, either self-exiled (Courant, Noether, Weyl) or dismissed (Landau).

12

French mathematics suffered deeply from its terrible loss of young men in the war, but it still had outstanding figures: Hadamard, Fréchet, Gaston Julia, Paul Levy, Borel, and Cartan, whose studies ranged from abstract spaces and function theory to group theory and probability. Paris remained the center, with vital research in physics represented by the Curies and Paul Langevin, thesis adviser to Louis de Broglie (1908). Results were often published in series of books, such as the *Actualités scientifiques et industrielles* (1929 and later) and the already mentioned *Mémorial.* Both the *Actualités* and the *Annales* of the Institut Henri Poincaré (1930 and later) contain studies in pure as well as applied mathematics and physics. The men of the older group taught and published long enough to

---

[31]This masculine title brought her no salary. But a *Lehrauftrag* provided her with a small income.

inspire the generation that, in 1940, started the Bourbaki enterprise.[32]

Another center, again with its own special character, was developing in Cambridge, England, breaking Great Britain's long insularity. Here again excellent mathematics was flanked by excellent physics: from 1919 Ernest Rutherford presided over the Cavendish Laboratory. There the modern age in analysis was represented by J. E. (John Edensor) Littlewood and G. H. (Godfrey Harold) Hardy. Hardy was from his student days in 1896 till his retirement in 1942 at Trinity College, except for a stay at Oxford from 1919 to 1931; Littlewood stayed at Cambridge from his student days until his retirement in 1950 (from 1910 also at Trinity), with only three years at Manchester. Hardy's *Course in Pure Mathematics* (1908) introduced into England in a rigorous fashion the then modern concepts of analysis (number, limit, function). Much of his further research—all in the "pure" mathematics that he extolled in his often quoted and controversial *A Mathematician's Apology* (1940)[33]—was achieved, from 1911 on, in collaboration with Littlewood, and in this work there are "Hardy-Littlewood" theorems in Fourier series, the Waring and Goldbach problems, Diophantine approximations, and the prime-number theorem. The "romantic incident" in Hardy's life (as Hardy called it) was his discovery of the Indian number-theoretical genius Srinivara Ramanujan from Madras, for whom Hardy made it possible to stay at Cambridge from 1917 to 1919, after which he returned to India to die in 1920 at the age of thirty-two. Collaboration between the two men led to many results, especially in *partitio numerorum*.

Hardy had been a student of Bertrand Russell, who, with Hardy, attracted many visitors to Cambridge, as did A. N. Whitehead, Russell's collaborator on the *Principia*. Close colleagues of Hardy were the analysts A. E. Besikovitch and E. C. Titchmarsh, the latter the author of a well-known *Theory of Functions* (1932). Between the pure mathematics of Hardy and the physics of Rutherford stood R. H. Fowler, one of whose students was P. A. M. Dirac. Both Dirac and Littlewood assisted Fowler in composing his standard *Statistical Mechanics* (1929).

All these men attracted students and visitors to Cambridge. Especially in the thirties, with Hardy back from Oxford, so many mathematicians, often refugees from fascism or revolution, went to Cambridge that that university replaced Göttingen as a great hub of mathematical life.

Edmund T. Whittaker, from 1912 to 1946 at Edinburgh, was a mathematical physicist (and Catholic philosopher), but obliged whole generations of students with the handsome "Whittaker-Watson" *Modern Analysis* (1915), written together with G. N. Watson, then at Cambridge. It is a fine exposition of the main transcendental complex functions. It has exercises, some pretty tough, as do also the Hardy and Titchmarsh books, an English tradition coming out of the old Tripos exams with their emphasis on problem-solving gymnastics, a tradition only gradually abandoned.

Cambridge-educated William Henry Young held several academic positions,

---

[32]See C. Boyer, *History of Mathematics* (1968), pp. 674–76.
[33]Reprinted 1967 with a foreword by C. P. Snow (Hardy had died in 1947).

one in Calcutta (1913–16). Independently he introduced (around 1902) the methods of the French school of Lebesgue and Baire into England, and worked on Fourier series and other fields of analysis. This was in collaboration with his wife, Grace Chisholm Young,[34] who had written her dissertation with Felix Klein on spherical trigonometry from the point of view of group theory (1895). They were the parents of Laurence Young, who has given us a very personal account of the Cambridge of the Hardy-Littlewood days, just as the Göttingen atmosphere has been described by Constance Reid.[35]

13

The October Revolution of 1917 made a powerful impact on the growth of Russian science, and mathematics shared in it. A strong tradition already existed, based on Lobačevskiĭ, M. V. Ostrogradskiĭ, and especially P. L. Čebyšev (Tschebycheff), the founder of the St. Petersburg school of mathematics, under whose inspiration A. A. Markov and A. M. Ljapunov received their training. Čebyšev was active in St. Petersburg from 1847 till his death in 1894, enriching many fields, from number theory (prime-number theorem), approximation problems, integration, and differential geometry, to kinematics and probability, with a keen sense for the relations between pure and applied domains. In probability he stressed sharp definitions, and led Markov (from 1886 to 1905 professor, till 1922 emeritus, at St. Petersburg) to the "Markov chains" of random variables (1906 and later). These chains have shown their importance in statistical physics, genetics, and economics; their theoretical foundations were strengthened by A. N. Kolmogorov.[36]

Ljapunov's researches were in the line of Laplace, both in his contributions to the theory of probabilities, where he generalized the fundamental limit theorem (1900–01), and also in his investigations into celestial mechanics.

To the St. Petersburg school also belongs G. F. Voronoĭ, after 1894 a professor at Warsaw (then under Russian rule), worth mentioning for his researches in number theory.

After the Revolution, Moscow became the capital of the Soviet Union. Here the Moscow school of mathematics was developed, mainly under the many-sided influence of N. N. Luzin, pupil of D. T. Egorov, who is known for a theorem on measurable functions (1911). Luzin went to Göttingen and Paris (1901, 1910), and taught at Moscow from 1914 till his death in 1953. He was among the first to apply measure theory to real functions, and he also especially investigated trigonometric series. In his seminars and through his lectures and textbooks he

---

[34]*Theory of Sets of Points* (1906).

[35]L. Young, *Mathematicians and Their Times* (Amsterdam, etc., 1981); C. Reid, *Hilbert* (New York, 1970); *Courant in Göttingen and New York* (New York, 1976). One of Young's early teachers was headmaster Edwin A. Abbott, whose fantasy of a two-dimensional world, *Flatland* (1884), was widely enjoyed, especially when Minkowski and Einstein introduced their four-dimensional world, a world that had to be popularized because of the mass interest in Einstein's theories during the 1920s and '30s.

[36]For the development of the Markov chains during these years, cf. M. Fréchet, *Recherches théoriques modernes sur le calcul des probabilités* (Paris, 1934).

educated whole generations of younger men in many fields of analysis, integration, and set theory. Sierpiński, who was to Poland what Luzin was to Moscow, was in close contact with Luzin.

Among the younger men influenced by Luzin were Paul S. Aleksandrov, A. Ja. Hinčin (Khintchin), P. S. Urysohn, A. N. Kolmogorov, P. A. Ljusternik, and L. S. Pontrjagin. Aleksandrov, with Pontrjagin and Urysohn (the latter of whom was drowned in 1924 at the age of twenty-six) established the Moscow topological school, which maintained steady contact with the West (Göttingen, Brouwer, Hausdorff). Several of these young men followed Luzin in his research on function theory. Typical of this whole area of activity is the care bestowed on the connections between theory and applications to mechanics, physics, and industry, an enterprise encouraged by the Soviet government. The study of probability remained alive; one of the best-known theoretical results is the set-theoretically based axiomatics of Kolmogorov: *Grundbegriffe der Wahrscheinlichkeitsrechnung* (1934).

The leading arithmetician was I. M. Vinogradov, first at Leningrad, and after 1934 at Moscow. His many contributions, influenced by Voronoï, and in many respects akin to those of Hardy and Littlewood, dealt with the position theory of numbers, the Waring and Goldbach problems. In this connection we think of Hilbert's eighth problem.

Geometry was represented by V. F. Kagan, first at Odessa, and after 1922 at Moscow. His earlier research was on the foundations of geometry, in the spirit of Hilbert, but at Moscow he applied himself to differential geometry and tensor calculus, to which he devoted an internationally known seminar, with its *Trudy* ("Transactions") published in 1933 and later. By this time, he was already an authority on Lobačevskiĭ.

A Ukrainian university of importance was Harkov (Kharkov), where Serge Bernstein taught from 1907 to 1933, after which he went to St. Petersburg and, in 1943, to Moscow. He had studied in Paris and Göttingen, wrote his thesis in Paris (1904), and in his further papers showed the influence both of Čebyšev (approximations, probability) and Weierstrass. Here again we find a deep interest in problems of applied mathematics (biology).

In 1919 Poland became an independent nation. We have seen that already in 1911 Sierpiński had laid the foundations of the Polish school of topology, which led to the *Fundamenta Mathematica*, the first journal exclusively devoted to one restricted subject area. Among Sierpiński's students we find K. Kuratowski and Alfred Tarski, the latter of whom was known as a logician and later went to Berkeley, California.[37] A second school developed under Stefan Banach at Lwów (then in Poland) and under Hugo Steinhaus at Lublin. In this school Banach made his great contributions to functional analysis, raising it from a study of the more special aspects due to Volterra and Hilbert to a comprehensive field. This contribution of Banach's was closely related to his introduction of the concept of normed linear spaces, in which the completely normed ones were named after him (1922

---

[37]K. Kuratowski, *A Half Century of Polish Mathematics: Remembrances and Reflections* (Oxford and Warsaw, 1980).

and later). Steinhaus paid much attention to applications, from probability to biology and engineering; he endeared himself to many people by his *Mathematical Snapshots*, a model of visual mathematics.[38] Banach and Steinhaus, from 1929 on, published the journal *Studia Mathematica*.

Polish mathematics was seriously affected by the Nazi occupation from 1939 to 1945. Many mathematicians who did not flee the country were killed. Banach and Steinhaus survived the horrors somehow, but Banach survived only by some months.

14

In Italy, the geometrical tradition was strong, and particularly that of algebraic geometry, nourished by A. Brill and Max Noether in Germany. Here we mention, apart from Segre, the publications and teachings of Guido Castelnuovo, Francesco Severi, and Federigo Enriques. Many of their results can be studied in Severi's books, notably in the German *Vorlesungen über algebraische Geometrie* (1921). This book and later studies deal with algebraic curves and varieties of higher dimension, Riemannian surfaces, and Abelian integrals. Enriques was, like Klein in Germany, also strongly concerned with educational questions, and his *Problèmes de la science et la logique* (1909), as well as his *Storia del pensiero scientifico* (1932, with Giorgio de Santillana, who later taught at MIT) shows Enriques' concern with the philosophy of science, in which he took a realist position opposed to current positivistic-idealistic trends.

These men all ended up at Rome and attracted many students and visitors, the more so since not only Volterra, but also Levi-Civita resided there (Volterra after 1900, Levi-Civita after 1918). Active in many fields besides tensors, Levi-Civita contributed to mechanics (the three-body problem), hydrodynamics (waves in canals), and relativity; both he and Cartan were welcome correspondents of Einstein. He was always ready to offer gracious guidance to younger men.

Fascism, more gradually than in Germany, also took its toll in Italy. Both Volterra and Levi-Civita lost their teaching positions. Others had to go abroad, like the physicist Enrico Fermi.

The entrance of the Netherlands into the modern capitalist world brought with it a renaissance of art and science, including physics (J. D. van der Waals and H. A. Lorentz), astronomy (J. C. Kapteyn), and mathematics. Lorentz's *Theory of Electrons* dates from 1909. The leading mathematicians were D. J. Korteweg at Amsterdam, J. C. Kluyver at Leiden, and P. H. Schoute at Groningen. Korteweg's main contributions were in mechanics and hydrodynamics, in which field we have a "theorem of Korteweg-DeVries"; Kluyver's papers were mostly in the spirit of Chasles and Hermite; Schoute, as we have seen, contributed to the field

---

[38]Crystallography, as well as considerations concerning the aesthetic values of mathematics, also led to aspects of visual mathematics. An influential book was *The Elements of Dynamic Symmetry* (1926), by the American painter and illustrator Jay Hambidge (Dover reprint, 1967). See also Hermann Weyl's *Symmetry* (1952), as well as G. D. Birkhoff's already mentioned *Aesthetic Measure* (1933, rev. 1961) and A. Speiser, *Theorie der Gruppen von endlicher Ordnung* (Berlin, 1923; 4th ed. Basel, 1956). These books have interesting illustrations. See also footnote 4, above.

of polytopes. Korteweg was the thesis adviser of L. E. J. Brouwer[39] and an editor of the *Oeuvres* of Huygens. Brouwer was also influenced by G. Mannoury, a self-taught mathematician, who introduced topology and foundation studies into the Netherlands and was also a founder of that type of semantics called *significa*. A younger generation includes J. A. Schouten, of tensor fame, who attracted many students; among his assistants was D. van Dantzig, active in many fields, from projective differential geometry and topology to, later on, mathematical statistics.[40] Bartel L. van der Waerden is a graduate of the University of Amsterdam (thesis: 1926). In the Nazi period Hans Freudenthal came to the Netherlands, and since 1946 has been at Utrecht.

In Hungary we mention F. Riesz, at Szeged (after 1946 at Budapest), who contributed to functional analysis and the study of linear operators in Hilbert space (we have mentioned the Riesz-Fischer theorem), and Lipot Fejér, from 1911 till his death in 1959 (with a short interruption) at Budapest, who did much of his research in harmonic analysis. Hans Hahn, at Vienna, was also mainly an analyst and studied real functions and abstract spaces.

Hahn was interested in the philosophy of mathematics and was instrumental in bringing Moritz Schlick to Vienna. Around Schlick, in the years after 1922, the so-called Vienna Circle was formed, consisting of mathematicians and other scientists interested in constructing an outlook on the world based on science "without metaphysics." To this group of "logical positivists" belonged the mathematicians Hahn, Gödel (who studied with Hahn), and Rudolf Carnap. Especially influential was Ludwig Wittgenstein. Carnap was the author of *Die logische Syntax der Sprache* (1934), Wittgenstein of the *Tractatus logico-philosophicus* (1922), each, in his own way, trying to build a road joining semantics, mathematical logic, and foundations of scientific inquiry. The circle had to disband with the advent of fascism, several of its members receiving influential positions abroad. Carnap went to Chicago, Gödel to Princeton, and Wittgenstein to Cambridge, England.[41]

In Scandinavia, we find T. A. Skolem in Norway, T. Carleman in Stockholm, and Harald Bohr (quasi-periodic functions) in Copenhagen; in Switzerland, Hurwitz and Minkowski at Basel (Minkowski from 1896 to 1902). From 1911 the monumental *Opera omnia* of Euler appeared under Swiss sponsorship.

Japan entered the modern period in mathematics during these years. We mention the algebraist Tejii Takagi and his pupils (Abelian fields, connection with Hilbert's twelfth problem). Differential geometry and tensors were cultivated by A. Kawaguchi and those he influenced. New schools of mathematics also began to develop in India, Canada, and Czechoslovakia.

---

[39]Brouwer's thesis, mentioned in Section 5, above, was originally so full of "philosophy" that Korteweg recommended leaving much of it out. See W. P. van Stigt, *HM* Vol. 6 (1979), pp. 385–404.

[40]See *Two Decades of Mathematics in the Netherlands, 1920–40*, E. M. J. Bertin, H. J. M. Bos, and A. W. Grootendorst, eds. (2 parts; Amsterdam, 1970). See also the article by Van der Corput, in the "Literature" section, below.

[41]The idealism in this approach has been subjected to much critique; see, e.g., M. Cornforth, *Marxism and the Linguistic Philosophy* (New York, 1965). Cornforth was one of the students who gathered around Wittgenstein at Cambridge, *c.* 1930.

15

Mathematics in the U.S.A. was brought up to world standards by men who had gone to study in Europe in the 1880s, especially in Germany. Among these were E. H. Moore, who, as we saw, founded his own school at Chicago, as well as William F. Osgood and Maxime Bôcher, both for many years at Harvard (1890–1933, 1891–1918, respectively). Osgood's *Lehrbuch der Funktionentheorie* (1907) was long a standard text because of its precision and form of exposition.

A younger generation included George D. Birkhoff at Harvard and Oswald Veblen at Princeton (1912–44, 1905–60, respectively), Veblen from 1932 to 1950 at the newly founded Institute for Advanced Study.[42] Birkhoff, after his success in 1913 with his proof of Poincaré's "last theorem" on the restricted three-body problem, continued to work in the spirit of Poincaré, enriching it with concepts of metric transitivity and, after 1931, with ergodic theorems. His attention went in many directions, including his theory of gravitation (1944, agreeing with Einstein on his special, but not on his general, theory of relativity). We have already mentioned his *Aesthetic Measure* (1933, richly illustrated).[43] His son is Garrett Birkhoff, who began his work in Boolean algebras (lattices) in the 1940s.

Veblen started in axiomatics, which set its stamp upon the two-volume *Projective Geometry* (1912, 1918), written with John W. Young, beginning in 1911, at Dartmouth.[44] His *Analysis Situs* (1922) opened a new way for a number of mathematicians and led to the Princeton School of Topology. From work in combinatory topology, the school passed, with J. W. Alexander and Solomon Lefschetz, to algebraic topology and homological algebra (which Lefschetz later introduced into Mexico). With J. H. C. Whitehead, Veblen wrote *The Foundations of Differential Geometry* (1932).

To U.S. mathematics of this period belongs the work of the older L. E. Dickson, with his studies on finite groups and impressive three-volume *History of the Theory of Numbers* (1919–23); the early work of John von Neumann, Hungarian-born, who after a lectorship at Göttingen, came to Princeton in 1930; and that of Norbert Wiener. To von Neumann's studies at that time belong group theory and Hilbert spaces, operators and ergodic theorems, with fundamental contributions to Hilbert's fifth problem. Von Neumann also devoted efforts to the mathematics of quantum mechanics and quantum thermodynamics. Wiener, after an early excursion into logic, launched out into Brownian motion and a generalized har-

---

[42]A pure research institution, privately endowed, founded in accordance with ideas outlined by the educator-scholar Abraham Flexner in his critical *Universities, American, British, German* (1930). It began with a School of Mathematics, whose early members included Veblen, von Neumann, and Marston Morse, as well as Einstein, Weyl, and other scientists from Nazi-dominated countries.

[43]On this subject see the bibliography by W. L. Schaaf, *Amer. Math. Monthly*, Vol. 55 (1951), pp. 157–77.

[44]The American geometer John W. Young (1879–1932) should not be confused with William H. Young, the British analyst (1863–1942) mentioned before. Then there is also the British clergyman Alfred Young (1873–1940), who, with J. H. Grace, his fellow student at Cambridge, was author of *The Algebra of Invariants* (1903), written in the spirit of Gordan and Frobenius.

monic analysis that involved Tauberian theorems,[45] studies leading to his formulation of cybernetics (1948). From 1919 till his death in 1964, Wiener taught at MIT.[46]

Other mathematicians of this period included H. Marston Morse, who continued Birkhoff's work in topology and the calculus of variations, and Marshall H. Stone, with his study of linear operators in Hilbert spaces and Boolean algebras.

Mathematics in the U.S.A. received an immense impetus through the immigration of eminent Europeans during the Nazi period. We have already mentioned several, among them Einstein, Weyl, and Emmy Noether. Others are E. Artin, R. Courant, G. Polya, H. Rademacher, V. Hurewicz, O. Neugebauer, André Weil, and R. von Mises. Tamarkin, at Brown University, had already come from Russia.[47]

16

The great era of the electronic computer came only after the Second World War, but the computer had a long and interesting prehistory, going as far back as Wilhelm Schickard, friend of Kepler (1623–24), Pascal, and Leibniz. In 1808 the weaver Joseph-Marie Jacquard invented a method to program a drawloom by means of punch cards. This idea was taken over by Charles Babbage for his "analytical engine" (1833). (Babbage was aided by Lady Ann Lovelace, daughter of Byron.) This engine, never finished, contained many ideas basic to any modern automatic computer: it could "store," perform calculations ("mill"), and control. But since the workings had to be entirely mechanical, only the electronics of the present age would have made it practical.

Between 1884 and 1890 Herman Hollerith, an American statistician engaged in the U.S. census of 1890, developed a system of mechanisms operating on punch cards, one for each person, every punch position representing a condition (profession, age, etc.). Konrad Zuse, a German, in 1934, improved on this system by taking up Leibniz's ideas on the use of the binary system.

Independent of this, Vannevar Bush, at MIT, in collaboration with Norbert Wiener, constructed in 1939 an analog computer for the evaluation of certain integrals and for solving some types of differential equations. In 1936, at Princeton, Alan M. Turing, a young Englishman, defined the "Turing machine," an abstract model of a possible logical machine, mentally constructed for such questions as Hilbert's decision problem. Later, after 1945 at Manchester in England, Turing applied his ideas to the construction of practical computers. Claude E. Shannon, then at MIT, took further steps in logical design for his information theory.

---

[45]Alfred Tauber (1866–1942), at Vienna, published certain integral conditions in a study on series (1897), used by Hardy and Littlewood and given the name of Tauberian theorems.

[46]The Massachusetts Institute of Technology, where collaboration between mathematicians, physicists, and electrical engineers, typically in the work of Norbert Wiener and others, such as Vannevar Bush, led to early work on computers.

[47]See further G. Birkhoff, "Some Leaders in American Mathematics," in *The Bicentennial Tribute to American Mathematics, 1776–1976*, D. Tarwater, ed. (Math. Assoc. of Amer., 1977), pp. 25–78.

The new era in working computers begins with the Mark I, started in 1937 at Harvard by Howard H. Aiken, with help from IBM (International Business Machines Corporation). Big industry began to take an interest. The Mark I thus had all the benefits of modern technology and finance, but though it contained magnetic devices it was still mainly mechanical. Improvements now came rapidly; the Mark II (1945–47) had all arithmetical and transfer operations done by electromagnetic relays. The first electronic computer, the ENIAC, was completed at the University of Pennsylvania in Philadelphia in 1946. All this was still university work, but in the 1950s computers became available for commercial purposes: the computer era had begun.

Literature

General surveys may be found in the already mentioned books by C. Boyer, M. Kline, N. Bourbaki, and H. Wussing and in the appendix by I. B. Pogrebysskiĭ to the Russian translation of this *Concise History* (also available in the German translation). Moreover, see:

Dubbey, J. M. *Development of Modern Mathematics.* New York, 1970.
Freudenthal, H. "The Implicit Philosophy of Mathematics Today." In *Contemporary Philosophy, a Survey,* R. Klibansky, ed. Florence, 1968, pp. 342–68.
Prasad, G. *Mathematical Research in the Last Twenty Years.* Berlin, 1923.
Weyl, H. "A Half Century of Mathematics." *Amer. Math. Monthly,* Vol. 58 (1951), pp. 523–83.

Special subjects, apart from the bio- and bibliographies in *DSB*, the short sketches in Meschkowski's *Lexikon,* and the already mentioned books by Black, Birkhoff, Kuratowski, Bertin, and Young, are taken up in the following writings:

Benacerraf, P., and Putnam, H. *Philosophy of Mathematics: Selected Readings.* Englewood Cliffs, N.J., 1964. (Essays by Carnap, von Neumann, Bernays, Gödel, Wittgenstein, and others.)
Bockstael, P. *Het Intuitionisme bÿ de Franse Wiskundigen.* Verh. Kon. Vlaamse Acad. Wet., 11 (1949), No. 32.
Bott, R. "Marston Morse and His Mathematical Works." *Bull. Amer. Math. Soc. (New Ser.),* Vol. 3 (1980), pp. 907–50.
Cahiers du séminaire d'histoire des mathématiques (1980–present). (Many papers on contemporary topics and authors.).
van Dalen, D. and Monna, A. F. *Sets and Integration. An Outline of the Development.* Groningen, 1973.
Dieudonné, J. *Cours de géométrie algébrique I.* Paris, 1974. (Gives history of this subject to past 1950.)
———. *History of Functional Analysis.* Amsterdam, 1981.
Felix, L. *The Modern Aspect of Mathematics.*
Goldstine, H. H. *The Computer from Pascal to von Neumann.* Princeton, N.J., 1970.
Grattan-Guinness, I. "On the Development of Logic Between the Two World Wars." *Amer. Math. Monthly,* Vol. 88 (1981), pp. 495–529.

Hawkins, J. *Lebesgue's Theory of Integration*. Madison, Wis., 1970.

Heins, S. J. *John von Neumann and Norbert Wiener.* Cambridge, Mass., 1980.

Le Lionnais, F. *Les grands courants de la pensée mathématique*. Paris, 1948; 2nd ed. augmentée, 1962. (A collection of essays by Borel, Fréchet, Denjoy, Weil, and others.)

On Luzin: *Uspekhi Matem. Nauk*, Vols. 6 (1951), 7 (1952), and 8 (1953).

Kennedy, H. *Life and Work of Giuseppe Peano*. Dordrecht and Boston, 1980.

Emmy Noether: A Tribute to Her Life and Work, J. W. Brewer and M. K. Smith, eds. New York and Basel, 1981.

Reid, C. *Hilbert*. Berlin, etc., 1970.

——. *Courant in Göttingen and New York*. New York, 1976.

Resnik, M. D. *Frege and the Philosophy of Mathematics*. Ithaca and London, 1980.

Van der Corput, J. G. "Wiskunde." In *Geestelijk Nederland 1920–1940*, K. F. Proost and J. Romein, eds. Amsterdam and Antwerp, [1947?], pp. 255–91. See also *The Development of Science in the Netherlands During the Last Half Century*. Leiden, 1930, pp. 44–51.

On Volterra: *Rendiconti Semin. Mat. e Fis. Milano*, Vol. 17 (1946), pp. 6–61.

Weil, A. "L'avenir des mathématiques." In Le Lionnais, above, pp. 307–20.

On N. Wiener: *Bull. Amer. Math. Soc.*, Vol. 72, No. 1, Part II, 1966 (145 pp.).

Index

Abacus, 81
Abbott, Edwin A., 210
Abel, Niels Henrik (1802–1829), 133, 151, 154, 155, 157, 158, 197, 212, 213
Abū Kāmil (c. 850–930), 73, 80
Abū-l-Wafa (940–997/8), 70
Académie des Sciences, 101, 102
Académie Française, 102
Achaemenids, 39
Achilles, 43, 112
Ackermann, Wilhelm (1896–1962), 202
Adélard of Bath (*fl.* 1120), 79
Agnesi, M. G. (1718–1799), 186
Ahrens, W., 199
Aiken, Howard H., 216
Akbar (1542–1605), 66
Albategnius = Al-Battānī (c. 850–929), 70
Alberti, Leone Battista (1404–1472), 84, 94
Albertus Magnus (c. 1208–1280), 91
Alcuin (735–804), 66, 78
Aleksandrov, Paul S., 211
Alexander, James Waddell (1888–1971), 199, 214
Alexander the Great (356–323 B.C.), 47
Al-Fazārī (d. c. 800), 67
Alfonsine tables, 73
Algorithmus = algorithm, 68, 81
Alhazen = Al Haitham (c. 965–1039), 72, 73
al-jabr = algebra, 68, 69, 81
Al-Karkhī = Al Karajī (d. c. 1029), 70, 73
Al-Kāshī (d. c. 1430), 72
Al-Khwārizmī (c. 780–c. 850), 68, 69, 73, 80
Almagest, 57, 66, 69
Al-Ma'mūn (786–833), 68
Al-Mansūr (c. 710–775), 68
Al-Zarqāli = Arzachel (c. 1029–1087), 73

Ampère, André Marie (1775–1836), 149
Anthoniszoon, Adriaen (c. 1543–1620), 73
Antonines (86–180), 56
apagoge, 40
Apollonius (c. 262–c. 190 B.C.), 48, 51, 54, 69, 83, 96, 99, 102, 108
Aquinas, St. Thomas (c. 1225–1274), 81–82
Arago, François (1786–1855), 136
Archibald, Raymond Claire (1875–1955), 206
Archimedes (c. 287–212 B.C.), 2, 44, 46, 48, 50–53, 55, 58–61, 69, 73, 83, 93–95, 100, 102, 118, 130
Archytas (*fl. c.* 400–c. 360 B.C.), 41, 44
Aristarchus (*fl.* 280–260 B.C.), 55
Aristotle (384–322 B.C.), 41–44, 67, 81, 96, 176
arithmos, 60
Aronhold, Siegfried Heinrich (1819–1884), 173, 176
Arrow, 43
Artin, Emil (1898–1962), 194, 200, 215
Āryabhata (d. 476), 66
Arzachel. See Al-Zarqāli.
Aśoka (273–232 B.C.), 31
At-Tusi. See Naṣīr al-dīn.
Augustine, St. (354–430), 81, 162

Babbage, Charles (1792–1871), 171, 215
Bacon, Francis (1561–1626), 95
Baire, Louis-René (1874–1932), 195, 210
Bakshāli manuscript, 67
Balzac, Honoré de (1799–1850), 152
Banach, Stefan (1892–1945), 195, 197, 211, 212
Barrow, Isaac (1630–1677), 100, 106, 111
Bartels, J. M., 170
Bayes, Thomas (d. 1763), 136
Beethoven, Ludwig van (1770–1827), 142

219

A CATALOG OF SELECTED
DOVER BOOKS
IN SCIENCE AND MATHEMATICS

Astronomy

CHARIOTS FOR APOLLO: The NASA History of Manned Lunar Spacecraft to 1969, Courtney G. Brooks, James M. Grimwood, and Loyd S. Swenson, Jr. This illustrated history by a trio of experts is the definitive reference on the Apollo spacecraft and lunar modules. It traces the vehicles' design, development, and operation in space. More than 100 photographs and illustrations. 576pp. 6 3/4 x 9 1/4. 0-486-46756-2

EXPLORING THE MOON THROUGH BINOCULARS AND SMALL TELESCOPES, Ernest H. Cherrington, Jr. Informative, profusely illustrated guide to locating and identifying craters, rills, seas, mountains, other lunar features. Newly revised and updated with special section of new photos. Over 100 photos and diagrams. 240pp. 8 1/4 x 11. 0-486-24491-1

WHERE NO MAN HAS GONE BEFORE: A History of NASA's Apollo Lunar Expeditions, William David Compton. Introduction by Paul Dickson. This official NASA history traces behind-the-scenes conflicts and cooperation between scientists and engineers. The first half concerns preparations for the Moon landings, and the second half documents the flights that followed Apollo 11. 1989 edition. 432pp. 7 x 10.
0-486-47888-2

APOLLO EXPEDITIONS TO THE MOON: The NASA History, Edited by Edgar M. Cortright. Official NASA publication marks the 40th anniversary of the first lunar landing and features essays by project participants recalling engineering and administrative challenges. Accessible, jargon-free accounts, highlighted by numerous illustrations. 336pp. 8 3/8 x 10 7/8. 0-486-47175-6

ON MARS: Exploration of the Red Planet, 1958-1978--The NASA History, Edward Clinton Ezell and Linda Neuman Ezell. NASA's official history chronicles the start of our explorations of our planetary neighbor. It recounts cooperation among government, industry, and academia, and it features dozens of photos from Viking cameras. 560pp. 6 3/4 x 9 1/4. 0-486-46757-0

ARISTARCHUS OF SAMOS: The Ancient Copernicus, Sir Thomas Heath. Heath's history of astronomy ranges from Homer and Hesiod to Aristarchus and includes quotes from numerous thinkers, compilers, and scholasticists from Thales and Anaximander through Pythagoras, Plato, Aristotle, and Heraclides. 34 figures. 448pp. 5 3/8 x 8 1/2.
0-486-43886-4

AN INTRODUCTION TO CELESTIAL MECHANICS, Forest Ray Moulton. Classic text still unsurpassed in presentation of fundamental principles. Covers rectilinear motion, central forces, problems of two and three bodies, much more. Includes over 200 problems, some with answers. 437pp. 5 3/8 x 8 1/2. 0-486-64687-4

BEYOND THE ATMOSPHERE: Early Years of Space Science, Homer E. Newell. This exciting survey is the work of a top NASA administrator who chronicles technological advances, the relationship of space science to general science, and the space program's social, political, and economic contexts. 528pp. 6 3/4 x 9 1/4.
0-486-47464-X

STAR LORE: Myths, Legends, and Facts, William Tyler Olcott. Captivating retellings of the origins and histories of ancient star groups include Pegasus, Ursa Major, Pleiades, signs of the zodiac, and other constellations. "Classic." – *Sky & Telescope.* 58 illustrations. 544pp. 5 3/8 x 8 1/2. 0-486-43581-4

A COMPLETE MANUAL OF AMATEUR ASTRONOMY: Tools and Techniques for Astronomical Observations, P. Clay Sherrod with Thomas L. Koed. Concise, highly readable book discusses the selection, set-up, and maintenance of a telescope; amateur studies of the sun; lunar topography and occultations; and more. 124 figures. 26 halftones. 37 tables. 335pp. 6 1/2 x 9 1/4. 0-486-42820-6

Chemistry

MOLECULAR COLLISION THEORY, M. S. Child. This high-level monograph offers an analytical treatment of classical scattering by a central force, quantum scattering by a central force, elastic scattering phase shifts, and semi-classical elastic scattering. 1974 edition. 310pp. 5 3/8 x 8 1/2. 0-486-69437-2

HANDBOOK OF COMPUTATIONAL QUANTUM CHEMISTRY, David B. Cook. This comprehensive text provides upper-level undergraduates and graduate students with an accessible introduction to the implementation of quantum ideas in molecular modeling, exploring practical applications alongside theoretical explanations. 1998 edition. 832pp. 5 3/8 x 8 1/2. 0-486-44307-8

RADIOACTIVE SUBSTANCES, Marie Curie. The celebrated scientist's thesis, which directly preceded her 1903 Nobel Prize, discusses establishing atomic character of radioactivity; extraction from pitchblende of polonium and radium; isolation of pure radium chloride; more. 96pp. 5 3/8 x 8 1/2. 0-486-42550-9

CHEMICAL MAGIC, Leonard A. Ford. Classic guide provides intriguing entertainment while elucidating sound scientific principles, with more than 100 unusual stunts: cold fire, dust explosions, a nylon rope trick, a disappearing beaker, much more. 128pp. 5 3/8 x 8 1/2. 0-486-67628-5

ALCHEMY, E. J. Holmyard. Classic study by noted authority covers 2,000 years of alchemical history: religious, mystical overtones; apparatus; signs, symbols, and secret terms; advent of scientific method, much more. Illustrated. 320pp. 5 3/8 x 8 1/2. 0-486-26298-7

CHEMICAL KINETICS AND REACTION DYNAMICS, Paul L. Houston. This text teaches the principles underlying modern chemical kinetics in a clear, direct fashion, using several examples to enhance basic understanding. Solutions to selected problems. 2001 edition. 352pp. 8 3/8 x 11. 0-486-45334-0

PROBLEMS AND SOLUTIONS IN QUANTUM CHEMISTRY AND PHYSICS, Charles S. Johnson and Lee G. Pedersen. Unusually varied problems, with detailed solutions, cover of quantum mechanics, wave mechanics, angular momentum, molecular spectroscopy, scattering theory, more. 280 problems, plus 139 supplementary exercises. 430pp. 6 1/2 x 9 1/4. 0-486-65236-X

ELEMENTS OF CHEMISTRY, Antoine Lavoisier. Monumental classic by the founder of modern chemistry features first explicit statement of law of conservation of matter in chemical change, and more. Facsimile reprint of original (1790) Kerr translation. 539pp. 5 3/8 x 8 1/2. 0-486-64624-6

MAGNETISM AND TRANSITION METAL COMPLEXES, F. E. Mabbs and D. J. Machin. A detailed view of the calculation methods involved in the magnetic properties of transition metal complexes, this volume offers sufficient background for original work in the field. 1973 edition. 240pp. 5 3/8 x 8 1/2. 0-486-46284-6

GENERAL CHEMISTRY, Linus Pauling. Revised third edition of classic first-year text by Nobel laureate. Atomic and molecular structure, quantum mechanics, statistical mechanics, thermodynamics correlated with descriptive chemistry. Problems. 992pp. 5 3/8 x 8 1/2. 0-486-65622-5

ELECTROLYTE SOLUTIONS: Second Revised Edition, R. A. Robinson and R. H. Stokes. Classic text deals primarily with measurement, interpretation of conductance, chemical potential, and diffusion in electrolyte solutions. Detailed theoretical interpretations, plus extensive tables of thermodynamic and transport properties. 1970 edition. 590pp. 5 3/8 x 8 1/2. 0-486-42225-9

Browse over 9,000 books at www.doverpublications.com

Engineering

FUNDAMENTALS OF ASTRODYNAMICS, Roger R. Bate, Donald D. Mueller, and Jerry E. White. Teaching text developed by U.S. Air Force Academy develops the basic two-body and n-body equations of motion; orbit determination; classical orbital elements, coordinate transformations; differential correction; more. 1971 edition. 455pp. 5 3/8 x 8 1/2.   0-486-60061-0

INTRODUCTION TO CONTINUUM MECHANICS FOR ENGINEERS: Revised Edition, Ray M. Bowen. This self-contained text introduces classical continuum models within a modern framework. Its numerous exercises illustrate the governing principles, linearizations, and other approximations that constitute classical continuum models. 2007 edition. 320pp. 6 1/8 x 9 1/4.   0-486-47460-7

ENGINEERING MECHANICS FOR STRUCTURES, Louis L. Bucciarelli. This text explores the mechanics of solids and statics as well as the strength of materials and elasticity theory. Its many design exercises encourage creative initiative and systems thinking. 2009 edition. 320pp. 6 1/8 x 9 1/4.   0-486-46855-0

FEEDBACK CONTROL THEORY, John C. Doyle, Bruce A. Francis and Allen R. Tannenbaum. This excellent introduction to feedback control system design offers a theoretical approach that captures the essential issues and can be applied to a wide range of practical problems. 1992 edition. 224pp. 6 1/2 x 9 1/4.   0-486-46933-6

THE FORCES OF MATTER, Michael Faraday. These lectures by a famous inventor offer an easy-to-understand introduction to the interactions of the universe's physical forces. Six essays explore gravitation, cohesion, chemical affinity, heat, magnetism, and electricity. 1993 edition. 96pp. 5 3/8 x 8 1/2.   0-486-47482-8

DYNAMICS, Lawrence E. Goodman and William H. Warner. Beginning engineering text introduces calculus of vectors, particle motion, dynamics of particle systems and plane rigid bodies, technical applications in plane motions, and more. Exercises and answers in every chapter. 619pp. 5 3/8 x 8 1/2.   0-486-42006-X

ADAPTIVE FILTERING PREDICTION AND CONTROL, Graham C. Goodwin and Kwai Sang Sin. This unified survey focuses on linear discrete-time systems and explores natural extensions to nonlinear systems. It emphasizes discrete-time systems, summarizing theoretical and practical aspects of a large class of adaptive algorithms. 1984 edition. 560pp. 6 1/2 x 9 1/4.   0-486-46932-8

INDUCTANCE CALCULATIONS, Frederick W. Grover. This authoritative reference enables the design of virtually every type of inductor. It features a single simple formula for each type of inductor, together with tables containing essential numerical factors. 1946 edition. 304pp. 5 3/8 x 8 1/2.   0-486-47440-2

THERMODYNAMICS: Foundations and Applications, Elias P. Gyftopoulos and Gian Paolo Beretta. Designed by two MIT professors, this authoritative text discusses basic concepts and applications in detail, emphasizing generality, definitions, and logical consistency. More than 300 solved problems cover realistic energy systems and processes. 800pp. 6 1/8 x 9 1/4.   0-486-43932-1

THE FINITE ELEMENT METHOD: Linear Static and Dynamic Finite Element Analysis, Thomas J. R. Hughes. Text for students without in-depth mathematical training, this text includes a comprehensive presentation and analysis of algorithms of time-dependent phenomena plus beam, plate, and shell theories. Solution guide available upon request. 672pp. 6 1/2 x 9 1/4.   0-486-41181-8

HELICOPTER THEORY, Wayne Johnson. Monumental engineering text covers vertical flight, forward flight, performance, mathematics of rotating systems, rotary wing dynamics and aerodynamics, aeroelasticity, stability and control, stall, noise, and more. 189 illustrations. 1980 edition. 1089pp. 5 5/8 x 8 1/4.   0-486-68230-7

MATHEMATICAL HANDBOOK FOR SCIENTISTS AND ENGINEERS: Definitions, Theorems, and Formulas for Reference and Review, Granino A. Korn and Theresa M. Korn. Convenient access to information from every area of mathematics: Fourier transforms, Z transforms, linear and nonlinear programming, calculus of variations, random-process theory, special functions, combinatorial analysis, game theory, much more. 1152pp. 5 3/8 x 8 1/2.   0-486-41147-8

A HEAT TRANSFER TEXTBOOK: Fourth Edition, John H. Lienhard V and John H. Lienhard IV. This introduction to heat and mass transfer for engineering students features worked examples and end-of-chapter exercises. Worked examples and end-of-chapter exercises appear throughout the book, along with well-drawn, illuminating figures. 768pp. 7 x 9 1/4.   0-486-47931-5

BASIC ELECTRICITY, U.S. Bureau of Naval Personnel. Originally a training course; best nontechnical coverage. Topics include batteries, circuits, conductors, AC and DC, inductance and capacitance, generators, motors, transformers, amplifiers, etc. Many questions with answers. 349 illustrations. 1969 edition. 448pp. 6 1/2 x 9 1/4.

    0-486-20973-3

BASIC ELECTRONICS, U.S. Bureau of Naval Personnel. Clear, well-illustrated introduction to electronic equipment covers numerous essential topics: electron tubes, semiconductors, electronic power supplies, tuned circuits, amplifiers, receivers, ranging and navigation systems, computers, antennas, more. 560 illustrations. 567pp. 6 1/2 x 9 1/4.   0-486-21076-6

BASIC WING AND AIRFOIL THEORY, Alan Pope. This self-contained treatment by a pioneer in the study of wind effects covers flow functions, airfoil construction and pressure distribution, finite and monoplane wings, and many other subjects. 1951 edition. 320pp. 5 3/8 x 8 1/2.   0-486-47188-8

SYNTHETIC FUELS, Ronald F. Probstein and R. Edwin Hicks. This unified presentation examines the methods and processes for converting coal, oil, shale, tar sands, and various forms of biomass into liquid, gaseous, and clean solid fuels. 1982 edition. 512pp. 6 1/8 x 9 1/4.   0-486-44977-7

THEORY OF ELASTIC STABILITY, Stephen P. Timoshenko and James M. Gere. Written by world-renowned authorities on mechanics, this classic ranges from theoretical explanations of 2- and 3-D stress and strain to practical applications such as torsion, bending, and thermal stress. 1961 edition. 560pp. 5 3/8 x 8 1/2.   0-486-47207-8

PRINCIPLES OF DIGITAL COMMUNICATION AND CODING, Andrew J. Viterbi and Jim K. Omura. This classic by two digital communications experts is geared toward students of communications theory and to designers of channels, links, terminals, modems, or networks used to transmit and receive digital messages. 1979 edition. 576pp. 6 1/8 x 9 1/4.   0-486-46901-8

LINEAR SYSTEM THEORY: The State Space Approach, Lotfi A. Zadeh and Charles A. Desoer. Written by two pioneers in the field, this exploration of the state space approach focuses on problems of stability and control, plus connections between this approach and classical techniques. 1963 edition. 656pp. 6 1/8 x 9 1/4.

    0-486-46663-9

Mathematics-Bestsellers

HANDBOOK OF MATHEMATICAL FUNCTIONS: with Formulas, Graphs, and Mathematical Tables, Edited by Milton Abramowitz and Irene A. Stegun. A classic resource for working with special functions, standard trig, and exponential logarithmic definitions and extensions, it features 29 sets of tables, some to as high as 20 places. 1046pp. 8 x 10 1/2.   0-486-61272-4

ABSTRACT AND CONCRETE CATEGORIES: The Joy of Cats, Jiri Adamek, Horst Herrlich, and George E. Strecker. This up-to-date introductory treatment employs category theory to explore the theory of structures. Its unique approach stresses concrete categories and presents a systematic view of factorization structures. Numerous examples. 1990 edition, updated 2004. 528pp. 6 1/8 x 9 1/4.   0-486-46934-4

MATHEMATICS: Its Content, Methods and Meaning, A. D. Aleksandrov, A. N. Kolmogorov, and M. A. Lavrent'ev. Major survey offers comprehensive, coherent discussions of analytic geometry, algebra, differential equations, calculus of variations, functions of a complex variable, prime numbers, linear and non-Euclidean geometry, topology, functional analysis, more. 1963 edition. 1120pp. 5 3/8 x 8 1/2.   0-486-40916-3

INTRODUCTION TO VECTORS AND TENSORS: Second Edition--Two Volumes Bound as One, Ray M. Bowen and C.-C. Wang. Convenient single-volume compilation of two texts offers both introduction and in-depth survey. Geared toward engineering and science students rather than mathematicians, it focuses on physics and engineering applications. 1976 edition. 560pp. 6 1/2 x 9 1/4.   0-486-46914-X

AN INTRODUCTION TO ORTHOGONAL POLYNOMIALS, Theodore S. Chihara. Concise introduction covers general elementary theory, including the representation theorem and distribution functions, continued fractions and chain sequences, the recurrence formula, special functions, and some specific systems. 1978 edition. 272pp. 5 3/8 x 8 1/2.
0-486-47929-3

ADVANCED MATHEMATICS FOR ENGINEERS AND SCIENTISTS, Paul DuChateau. This primary text and supplemental reference focuses on linear algebra, calculus, and ordinary differential equations. Additional topics include partial differential equations and approximation methods. Includes solved problems. 1992 edition. 400pp. 7 1/2 x 9 1/4.   0-486-47930-7

PARTIAL DIFFERENTIAL EQUATIONS FOR SCIENTISTS AND ENGINEERS, Stanley J. Farlow. Practical text shows how to formulate and solve partial differential equations. Coverage of diffusion-type problems, hyperbolic-type problems, elliptic-type problems, numerical and approximate methods. Solution guide available upon request. 1982 edition. 414pp. 6 1/8 x 9 1/4.   0-486-67620-X

VARIATIONAL PRINCIPLES AND FREE-BOUNDARY PROBLEMS, Avner Friedman. Advanced graduate-level text examines variational methods in partial differential equations and illustrates their applications to free-boundary problems. Features detailed statements of standard theory of elliptic and parabolic operators. 1982 edition. 720pp. 6 1/8 x 9 1/4.   0-486-47853-X

LINEAR ANALYSIS AND REPRESENTATION THEORY, Steven A. Gaal. Unified treatment covers topics from the theory of operators and operator algebras on Hilbert spaces; integration and representation theory for topological groups; and the theory of Lie algebras, Lie groups, and transform groups. 1973 edition. 704pp. 6 1/8 x 9 1/4.
0-486-47851-3

Browse over 9,000 books at www.doverpublications.com

A SURVEY OF INDUSTRIAL MATHEMATICS, Charles R. MacCluer. Students learn how to solve problems they'll encounter in their professional lives with this concise single-volume treatment. It employs MATLAB and other strategies to explore typical industrial problems. 2000 edition. 384pp. 5 3/8 x 8 1/2. 0-486-47702-9

NUMBER SYSTEMS AND THE FOUNDATIONS OF ANALYSIS, Elliott Mendelson. Geared toward undergraduate and beginning graduate students, this study explores natural numbers, integers, rational numbers, real numbers, and complex numbers. Numerous exercises and appendixes supplement the text. 1973 edition. 368pp. 5 3/8 x 8 1/2. 0-486-45792-3

A FIRST LOOK AT NUMERICAL FUNCTIONAL ANALYSIS, W. W. Sawyer. Text by renowned educator shows how problems in numerical analysis lead to concepts of functional analysis. Topics include Banach and Hilbert spaces, contraction mappings, convergence, differentiation and integration, and Euclidean space. 1978 edition. 208pp. 5 3/8 x 8 1/2. 0-486-47882-3

FRACTALS, CHAOS, POWER LAWS: Minutes from an Infinite Paradise, Manfred Schroeder. A fascinating exploration of the connections between chaos theory, physics, biology, and mathematics, this book abounds in award-winning computer graphics, optical illusions, and games that clarify memorable insights into self-similarity. 1992 edition. 448pp. 6 1/8 x 9 1/4. 0-486-47204-3

SET THEORY AND THE CONTINUUM PROBLEM, Raymond M. Smullyan and Melvin Fitting. A lucid, elegant, and complete survey of set theory, this three-part treatment explores axiomatic set theory, the consistency of the continuum hypothesis, and forcing and independence results. 1996 edition. 336pp. 6 x 9. 0-486-47484-4

DYNAMICAL SYSTEMS, Shlomo Sternberg. A pioneer in the field of dynamical systems discusses one-dimensional dynamics, differential equations, random walks, iterated function systems, symbolic dynamics, and Markov chains. Supplementary materials include PowerPoint slides and MATLAB exercises. 2010 edition. 272pp. 6 1/8 x 9 1/4. 0-486-47705-3

ORDINARY DIFFERENTIAL EQUATIONS, Morris Tenenbaum and Harry Pollard. Skillfully organized introductory text examines origin of differential equations, then defines basic terms and outlines general solution of a differential equation. Explores integrating factors; dilution and accretion problems; Laplace Transforms; Newton's Interpolation Formulas, more. 818pp. 5 3/8 x 8 1/2. 0-486-64940-7

MATROID THEORY, D. J. A. Welsh. Text by a noted expert describes standard examples and investigation results, using elementary proofs to develop basic matroid properties before advancing to a more sophisticated treatment. Includes numerous exercises. 1976 edition. 448pp. 5 3/8 x 8 1/2. 0-486-47439-9

THE CONCEPT OF A RIEMANN SURFACE, Hermann Weyl. This classic on the general history of functions combines function theory and geometry, forming the basis of the modern approach to analysis, geometry, and topology. 1955 edition. 208pp. 5 3/8 x 8 1/2. 0-486-47004-0

THE LAPLACE TRANSFORM, David Vernon Widder. This volume focuses on the Laplace and Stieltjes transforms, offering a highly theoretical treatment. Topics include fundamental formulas, the moment problem, monotonic functions, and Tauberian theorems. 1941 edition. 416pp. 5 3/8 x 8 1/2. 0-486-47755-X

Browse over 9,000 books at www.doverpublications.com

Mathematics–Logic and Problem Solving

PERPLEXING PUZZLES AND TANTALIZING TEASERS, Martin Gardner. Ninety-three riddles, mazes, illusions, tricky questions, word and picture puzzles, and other challenges offer hours of entertainment for youngsters. Filled with rib-tickling drawings. Solutions. 224pp. 5 3/8 x 8 1/2. 0-486-25637-5

MY BEST MATHEMATICAL AND LOGIC PUZZLES, Martin Gardner. The noted expert selects 70 of his favorite "short" puzzles. Includes The Returning Explorer, The Mutilated Chessboard, Scrambled Box Tops, and dozens more. Complete solutions included. 96pp. 5 3/8 x 8 1/2. 0-486-28152-3

THE LADY OR THE TIGER?: and Other Logic Puzzles, Raymond M. Smullyan. Created by a renowned puzzle master, these whimsically themed challenges involve paradoxes about probability, time, and change; metapuzzles; and self-referentiality. Nineteen chapters advance in difficulty from relatively simple to highly complex. 1982 edition. 240pp. 5 3/8 x 8 1/2. 0-486-47027-X

SATAN, CANTOR AND INFINITY: Mind-Boggling Puzzles, Raymond M. Smullyan. A renowned mathematician tells stories of knights and knaves in an entertaining look at the logical precepts behind infinity, probability, time, and change. Requires a strong background in mathematics. Complete solutions. 288pp. 5 3/8 x 8 1/2.
0-486-47036-9

THE RED BOOK OF MATHEMATICAL PROBLEMS, Kenneth S. Williams and Kenneth Hardy. Handy compilation of 100 practice problems, hints and solutions indispensable for students preparing for the William Lowell Putnam and other mathematical competitions. Preface to the First Edition. Sources. 1988 edition. 192pp. 5 3/8 x 8 1/2. 0-486-69415-1

KING ARTHUR IN SEARCH OF HIS DOG AND OTHER CURIOUS PUZZLES, Raymond M. Smullyan. This fanciful, original collection for readers of all ages features arithmetic puzzles, logic problems related to crime detection, and logic and arithmetic puzzles involving King Arthur and his Dogs of the Round Table. 160pp. 5 3/8 x 8 1/2.
0-486-47435-6

UNDECIDABLE THEORIES: Studies in Logic and the Foundation of Mathematics, Alfred Tarski in collaboration with Andrzej Mostowski and Raphael M. Robinson. This well-known book by the famed logician consists of three treatises: "A General Method in Proofs of Undecidability," "Undecidability and Essential Undecidability in Mathematics," and "Undecidability of the Elementary Theory of Groups." 1953 edition. 112pp. 5 3/8 x 8 1/2. 0-486-47703-7

LOGIC FOR MATHEMATICIANS, J. Barkley Rosser. Examination of essential topics and theorems assumes no background in logic. "Undoubtedly a major addition to the literature of mathematical logic." – Bulletin of the American Mathematical Society. 1978 edition. 592pp. 6 1/8 x 9 1/4. 0-486-46898-4

INTRODUCTION TO PROOF IN ABSTRACT MATHEMATICS, Andrew Wohlgemuth. This undergraduate text teaches students what constitutes an acceptable proof, and it develops their ability to do proofs of routine problems as well as those requiring creative insights. 1990 edition. 384pp. 6 1/2 x 9 1/4. 0-486-47854-8

FIRST COURSE IN MATHEMATICAL LOGIC, Patrick Suppes and Shirley Hill. Rigorous introduction is simple enough in presentation and context for wide range of students. Symbolizing sentences; logical inference; truth and validity; truth tables; terms, predicates, universal quantifiers; universal specification and laws of identity; more. 288pp. 5 3/8 x 8 1/2. 0-486-42259-3

Mathematics–Algebra and Calculus

VECTOR CALCULUS, Peter Baxandall and Hans Liebeck. This introductory text offers a rigorous, comprehensive treatment. Classical theorems of vector calculus are amply illustrated with figures, worked examples, physical applications, and exercises with hints and answers. 1986 edition. 560pp. 5 3/8 x 8 1/2.   0-486-46620-5

ADVANCED CALCULUS: An Introduction to Classical Analysis, Louis Brand. A course in analysis that focuses on the functions of a real variable, this text introduces the basic concepts in their simplest setting and illustrates its teachings with numerous examples, theorems, and proofs. 1955 edition. 592pp. 5 3/8 x 8 1/2.   0-486-44548-8

ADVANCED CALCULUS, Avner Friedman. Intended for students who have already completed a one-year course in elementary calculus, this two-part treatment advances from functions of one variable to those of several variables. Solutions. 1971 edition. 432pp. 5 3/8 x 8 1/2.   0-486-45795-8

METHODS OF MATHEMATICS APPLIED TO CALCULUS, PROBABILITY, AND STATISTICS, Richard W. Hamming. This 4-part treatment begins with algebra and analytic geometry and proceeds to an exploration of the calculus of algebraic functions and transcendental functions and applications. 1985 edition. Includes 310 figures and 18 tables. 880pp. 6 1/2 x 9 1/4.   0-486-43945-3

BASIC ALGEBRA I: Second Edition, Nathan Jacobson. A classic text and standard reference for a generation, this volume covers all undergraduate algebra topics, including groups, rings, modules, Galois theory, polynomials, linear algebra, and associative algebra. 1985 edition. 528pp. 6 1/8 x 9 1/4.   0-486-47189-6

BASIC ALGEBRA II: Second Edition, Nathan Jacobson. This classic text and standard reference comprises all subjects of a first-year graduate-level course, including in-depth coverage of groups and polynomials and extensive use of categories and functors. 1989 edition. 704pp. 6 1/8 x 9 1/4.   0-486-47187-X

CALCULUS: An Intuitive and Physical Approach (Second Edition), Morris Kline. Application-oriented introduction relates the subject as closely as possible to science with explorations of the derivative; differentiation and integration of the powers of x; theorems on differentiation, antidifferentiation; the chain rule; trigonometric functions; more. Examples. 1967 edition. 960pp. 6 1/2 x 9 1/4.   0-486-40453-6

ABSTRACT ALGEBRA AND SOLUTION BY RADICALS, John E. Maxfield and Margaret W. Maxfield. Accessible advanced undergraduate-level text starts with groups, rings, fields, and polynomials and advances to Galois theory, radicals and roots of unity, and solution by radicals. Numerous examples, illustrations, exercises, appendixes. 1971 edition. 224pp. 6 1/8 x 9 1/4.   0-486-47723-1

AN INTRODUCTION TO THE THEORY OF LINEAR SPACES, Georgi E. Shilov. Translated by Richard A. Silverman. Introductory treatment offers a clear exposition of algebra, geometry, and analysis as parts of an integrated whole rather than separate subjects. Numerous examples illustrate many different fields, and problems include hints or answers. 1961 edition. 320pp. 5 3/8 x 8 1/2.   0-486-63070-6

LINEAR ALGEBRA, Georgi E. Shilov. Covers determinants, linear spaces, systems of linear equations, linear functions of a vector argument, coordinate transformations, the canonical form of the matrix of a linear operator, bilinear and quadratic forms, and more. 387pp. 5 3/8 x 8 1/2.   0-486-63518-X

Mathematics-Probability and Statistics

BASIC PROBABILITY THEORY, Robert B. Ash. This text emphasizes the probabilistic way of thinking, rather than measure-theoretic concepts. Geared toward advanced undergraduates and graduate students, it features solutions to some of the problems. 1970 edition. 352pp. 5 3/8 x 8 1/2.   0-486-46628-0

PRINCIPLES OF STATISTICS, M. G. Bulmer. Concise description of classical statistics, from basic dice probabilities to modern regression analysis. Equal stress on theory and applications. Moderate difficulty; only basic calculus required. Includes problems with answers. 252pp. 5 5/8 x 8 1/4.   0-486-63760-3

OUTLINE OF BASIC STATISTICS: Dictionary and Formulas, John E. Freund and Frank J. Williams. Handy guide includes a 70-page outline of essential statistical formulas covering grouped and ungrouped data, finite populations, probability, and more, plus over 1,000 clear, concise definitions of statistical terms. 1966 edition. 208pp. 5 3/8 x 8 1/2.   0-486-47769-X

GOOD THINKING: The Foundations of Probability and Its Applications, Irving J. Good. This in-depth treatment of probability theory by a famous British statistician explores Keynesian principles and surveys such topics as Bayesian rationality, corroboration, hypothesis testing, and mathematical tools for induction and simplicity. 1983 edition. 352pp. 5 3/8 x 8 1/2.   0-486-47438-0

INTRODUCTION TO PROBABILITY THEORY WITH CONTEMPORARY APPLICATIONS, Lester L. Helms. Extensive discussions and clear examples, written in plain language, expose students to the rules and methods of probability. Exercises foster problem-solving skills, and all problems feature step-by-step solutions. 1997 edition. 368pp. 6 1/2 x 9 1/4.   0-486-47418-6

CHANCE, LUCK, AND STATISTICS, Horace C. Levinson. In simple, non-technical language, this volume explores the fundamentals governing chance and applies them to sports, government, and business. "Clear and lively ... remarkably accurate." – Scientific Monthly. 384pp. 5 3/8 x 8 1/2.   0-486-41997-5

FIFTY CHALLENGING PROBLEMS IN PROBABILITY WITH SOLUTIONS, Frederick Mosteller. Remarkable puzzlers, graded in difficulty, illustrate elementary and advanced aspects of probability. These problems were selected for originality, general interest, or because they demonstrate valuable techniques. Also includes detailed solutions. 88pp. 5 3/8 x 8 1/2.   0-486-65355-2

EXPERIMENTAL STATISTICS, Mary Gibbons Natrella. A handbook for those seeking engineering information and quantitative data for designing, developing, constructing, and testing equipment. Covers the planning of experiments, the analyzing of extreme-value data; and more. 1966 edition. Index. Includes 52 figures and 76 tables. 560pp. 8 3/8 x 11.   0-486-43937-2

STOCHASTIC MODELING: Analysis and Simulation, Barry L. Nelson. Coherent introduction to techniques also offers a guide to the mathematical, numerical, and simulation tools of systems analysis. Includes formulation of models, analysis, and interpretation of results. 1995 edition. 336pp. 6 1/8 x 9 1/4.   0-486-47770-3

INTRODUCTION TO BIOSTATISTICS: Second Edition, Robert R. Sokal and F. James Rohlf. Suitable for undergraduates with a minimal background in mathematics, this introduction ranges from descriptive statistics to fundamental distributions and the testing of hypotheses. Includes numerous worked-out problems and examples. 1987 edition. 384pp. 6 1/8 x 9 1/4.   0-486-46961-1

Mathematics–Geometry and Topology

PROBLEMS AND SOLUTIONS IN EUCLIDEAN GEOMETRY, M. N. Aref and William Wernick. Based on classical principles, this book is intended for a second course in Euclidean geometry and can be used as a refresher. More than 200 problems include hints and solutions. 1968 edition. 272pp. 5 3/8 x 8 1/2. 0-486-47720-7

TOPOLOGY OF 3-MANIFOLDS AND RELATED TOPICS, Edited by M. K. Fort, Jr. With a New Introduction by Daniel Silver. Summaries and full reports from a 1961 conference discuss decompositions and subsets of 3-space; n-manifolds; knot theory; the Poincaré conjecture; and periodic maps and isotopies. Familiarity with algebraic topology required. 1962 edition. 272pp. 6 1/8 x 9 1/4. 0-486-47753-3

POINT SET TOPOLOGY, Steven A. Gaal. Suitable for a complete course in topology, this text also functions as a self-contained treatment for independent study. Additional enrichment materials make it equally valuable as a reference. 1964 edition. 336pp. 5 3/8 x 8 1/2. 0-486-47222-1

INVITATION TO GEOMETRY, Z. A. Melzak. Intended for students of many different backgrounds with only a modest knowledge of mathematics, this text features self-contained chapters that can be adapted to several types of geometry courses. 1983 edition. 240pp. 5 3/8 x 8 1/2. 0-486-46626-4

TOPOLOGY AND GEOMETRY FOR PHYSICISTS, Charles Nash and Siddhartha Sen. Written by physicists for physics students, this text assumes no detailed background in topology or geometry. Topics include differential forms, homotopy, homology, cohomology, fiber bundles, connection and covariant derivatives, and Morse theory. 1983 edition. 320pp. 5 3/8 x 8 1/2. 0-486-47852-1

BEYOND GEOMETRY: Classic Papers from Riemann to Einstein, Edited with an Introduction and Notes by Peter Pesic. This is the only English-language collection of these 8 accessible essays. They trace seminal ideas about the foundations of geometry that led to Einstein's general theory of relativity. 224pp. 6 1/8 x 9 1/4. 0-486-45350-2

GEOMETRY FROM EUCLID TO KNOTS, Saul Stahl. This text provides a historical perspective on plane geometry and covers non-neutral Euclidean geometry, circles and regular polygons, projective geometry, symmetries, inversions, informal topology, and more. Includes 1,000 practice problems. Solutions available. 2003 edition. 480pp. 6 1/8 x 9 1/4. 0-486-47459-3

TOPOLOGICAL VECTOR SPACES, DISTRIBUTIONS AND KERNELS, François Trèves. Extending beyond the boundaries of Hilbert and Banach space theory, this text focuses on key aspects of functional analysis, particularly in regard to solving partial differential equations. 1967 edition. 592pp. 5 3/8 x 8 1/2. 0-486-45352-9

INTRODUCTION TO PROJECTIVE GEOMETRY, C. R. Wylie, Jr. This introductory volume offers strong reinforcement for its teachings, with detailed examples and numerous theorems, proofs, and exercises, plus complete answers to all odd-numbered end-of-chapter problems. 1970 edition. 576pp. 6 1/8 x 9 1/4. 0-486-46895-X

FOUNDATIONS OF GEOMETRY, C. R. Wylie, Jr. Geared toward students preparing to teach high school mathematics, this text explores the principles of Euclidean and non-Euclidean geometry and covers both generalities and specifics of the axiomatic method. 1964 edition. 352pp. 6 x 9. 0-486-47214-0